T0221192

Jane Austen, Game Theorist

Jane Austen, Game Theorist

Michael Suk-Young Chwe

PRINCETON UNIVERSITY PRESS

PRINCETON AND OXFORD

Copyright © 2013 by Princeton University Press

Published by Princeton University Press, 41 William Street, Princeton, New Jersey 08540

In the United Kingdom: Princeton University Press, 6 Oxford Street, Woodstock, Oxfordshire OX20 1TW

press.princeton.edu

Library of Congress Cataloging-in-Publication Data

Chwe, Michael Suk-Young, 1965–
Jane Austen, game theorist / Michael Suk-Young Chwe.
  pages cm
Includes bibliographical references and index.
ISBN 978-0-691-15576-0 (hardcover : acid-free paper)
1. Austen, Jane, 1775–1817—Criticism and interpretation.
2. Austen, Jane, 1775–1817—Knowledge—Social life and customs.
3. Game theory in literature. 4. Game theory—Social aspects.
5. Rational choice theory—Social aspects. I. Title.
  PR4038.G36C49 2013
  823'.7—dc23
                        2012041510

British Library Cataloging-in-Publication Data is available

This book has been composed in Sabon

Printed on acid-free paper ∞

Typeset by S R Nova Pvt Ltd, Bangalore, India
Printed in the United States of America

10 9 8 7 6 5 4 3

To my sister Lana

Contents

Contents

CHAPTER FOURTEEN
Concluding Remarks

Preface

THE IDEA for this book started when I found *Flossie and the Fox* (McKissack 1986) for my children at a garage sale. For years I used the story of Flossie as an example in my graduate game theory classes but never found a place for it in my writing. The opportunity came when I was asked to prepare a paper for a conference on "Rational Choice Theory and the Humanities." I found similar folktales and began to notice "folk game theory" in movies I watched together with my children. Watching Jane Austen adaptations led to reading her novels. Thus this book arose out of experiences with my children Hanyu and Hana. Now as they are almost grown I hope that they will still want to read books and watch movies with their father.

The "Rational Choice Theory and the Humanities" conference was held at Stanford University in April 2005, and I am indebted to the organizer, David Palumbo-Liu, and conference participants. Some of the material in the paper I wrote for the conference (Chwe 2009) appears again here. I am also indebted to participants in presentations I gave in December 2005 at the National Taiwan University, in April 2010 at the Juan March Institute and the UCLA Marschak Colloquium, in May 2011 at Yale University, and in June 2011 at the University of Oxford and the Stockholm School of Economics. Discussions continue at janeaustengametheorist.com.

Writing a book invariably exposes one to undeserved generosity. More than once, I have drafted what seems like a delightfully original phrase only to discover it in an email received earlier from a friend. The term "folk game theory" has also been independently coined in lectures by Vince Crawford and in a recent paper by Crawford, Costa-Gomes, and Iriberri (2010). Properly acknowledging the contributions of my friends and colleagues is almost impossible, but I will try. Specifically (in reverse alphabetical order), Guenter Treitel, Laura Rosenthal, Dick Rosecrance, Anne Mellor, Avinash Dixit, Vince Crawford, Tyler Cowen, Steve Brams, and Pippa Abston read the entire first draft and offered very helpful comments. Peyton Young, Giulia Sissa, Ignacio Sanchez-Cuenca, Valeria Pizzini-Gambetta, Rohit Parikh, Russ Mardon, and Neal Beck gave me great suggestions. The comments of anonymous referees improved the book, especially its overall organization, a lot. I am indebted. Chuck Myers and Peter Dougherty at Princeton University Press have always been great. Linda Truilo's care as copyeditor is very much appreciated.

I wrote this book while my wife and I were both teaching at UCLA, where we landed after moving three times in search of two jobs at the same place. We are grateful to this university for its research environment and also simply for making our family life possible. I will always look back on this period of my life, with our children growing up among many loving families in Santa Monica, with the greatest affection.

Finally, I would like to thank my own family: my wife and my children, my parents, and my brothers and my sister. If I could, I would dedicate everything I do to them, not just books.

Thank you for reading, even if these words appear on some sort of device with an off switch. Real books could come into your life serendipitously, from garage sales for example. We read them. We wrote them. We loved them.

Abbreviations

The following abbreviations are used for Jane Austen's novels.

E	*Emma*
MP	*Mansfield Park*
NA	*Northanger Abbey*
P	*Persuasion*
PP	*Pride and Prejudice*
SS	*Sense and Sensibility*

Jane Austen, Game Theorist

CHAPTER ONE

The Argument

NOTHING IS MORE human than being curious about other humans. Why do people do what they do? The social sciences have answered this question in increasingly theoretical and specialized ways. One of the most popular and influential in the past fifty years, at least in economics and political science, has been game theory. However, in this book I argue that Jane Austen systematically explored the core ideas of game theory in her six novels, roughly two hundred years ago.

Austen is not just singularly insightful but relentlessly theoretical. Austen starts with the basic concepts of choice (a person does what she does because she chooses to) and preferences (a person chooses according to her preferences). Strategic thinking, what Austen calls "penetration," is game theory's central concept: when choosing an action, a person thinks about how others will act. Austen analyzes these foundational concepts in examples too numerous and systematic to be considered incidental. Austen then considers how strategic thinking relates to other explanations of human action, such as those involving emotions, habits, rules, social factors, and ideology. Austen also carefully distinguishes strategic thinking from other concepts often confused with it, such as selfishness and economism, and even discusses the disadvantages of strategic thinking. Finally, Austen explores new applications, arguing, for example, that strategizing together in a partnership is the surest foundation for intimate relationships.

Given the breadth and ambition of her discussion, I argue that exploring strategic thinking, theoretically and not just for practical advantage, is Austen's explicit intention. Austen is a theoretician of strategic thinking, in her own words, an "imaginist." Austen's novels do not simply provide "case material" for the game theorist to analyze, but are themselves an ambitious theoretical project, with insights not yet superseded by modern social science.

In her ambition, Austen is singular but not alone. For example, African American folktales celebrate the clever manipulation of others, and I argue that their strategic legacy informed the tactics of the U.S. civil rights movement. Just as folk medicine healed people long before academic medicine, "folk game theory" expertly analyzed strategic situations long before game theory became an academic specialty. For example, the tale of Flossie and the Fox shows how pretending to be naive can deter attackers, a theory of deterrence at least as sophisticated as those in social science today. Folk game theory contains wisdom that can be explored by social science just as traditional folk remedies are investigated by modern medicine. Game theory should thus embrace Austen, African American folktellers, and the world's many folk game theory traditions as true scientific predecessors.

The connection between Austen's novels, among the most widely beloved in the English language, and game theory, which can be quite mathematical, might seem unlikely. Austen's novels are discerning and sensitive, whereas game theory is often seen as reductive and technical, originating out of a Cold War military-industrial "think tank." But since both Austen and game theory build a theory of human behavior based upon strategic thinking, it is not surprising that they develop the same concepts even as they consider different applications. Strategic thinking can reach a surprising level of virtuosity, but people actually do it all the time (for example, I hide the cookies because I know that otherwise you will eat them all). A theory based on strategic thinking is, of course, not the only theory of human behavior or always the most relevant, but it is useful and "universal" enough to have developed independently, in quite different historical contexts.

Why should we care about Austen's place in the history of game theory? The most obvious trend in the language of the social sciences over the past fifty years is a greater use of mathematics. A large part of this trend is the growth of game theory and its intellectual predecessor, rational choice theory. This growth is indeed one of the broadest developments in the social sciences in the past fifty years, significant enough to have broader social as well as scholarly implications; for example, some claim that the 2008 global financial crisis was caused in part by rationality assumptions in economics and finance (for example Stiglitz 2010; see also MacKenzie 2006).

Recognizing Austen as a game theorist helps us see how game theory has more diverse and subversive historical roots. Austen and African American folktellers speak as outsiders: women dependent on men, and slaves struggling for autonomy. They build a theory of strategic thinking not to better chase a Soviet submarine but to survive. The powerful can of course use game theory, but game theory develops distinctively among the subordinate and oppressed, people for whom making exactly the right strategic move in the right situation can have enormous consequences: women who might gain husbands, and slaves who might gain freedom. The dominant have less need for game theory because from their point of view, everyone else is already doing what they are supposed to. Game theory is not necessarily a hegemonic Cold War discourse but one of the original "weapons of the weak" (as in Scott

1985). By recovering a "people's history of game theory" (as in Zinn 2003) we enlarge its potential future.

Understanding Austen's six novels as a systematic research project also allows us to interpret many details not often examined. For example, why do Austen's Jane Fairfax and Mr. John Knightley discuss whether the reliability of postal service workers is due to interest or habit? When Emma Woodhouse paints a portrait of Harriet Smith and Mr. Elton adores it, why does Emma think that Mr. Elton is in love with the painting's subject, not its creator? Why is Fanny Price grateful not to have to choose between wearing Edmund Bertram's chain or Mary Crawford's necklace, but then decides to wear both? When they meet for the first time, why does Mrs. Croft ask Anne Elliot whether she has heard that Mrs. Croft's brother has married, without specifying which brother? There is, of course, an immense literature on Austen, and I cannot claim the primacy of my own reading. Still, a strategic sensibility can help generate and answer questions like these.

Recognizing Austen as a game theorist is worthwhile not only for the sake of intellectual genealogy. Anyone interested in human behavior should read Austen because her research program has results.

Austen makes particular advances in a topic not yet taken up by modern game theory: the conspicuous absence of strategic thinking, what I call "cluelessness." Even though strategic thinking is a basic human skill, often people do not apply it and even actively resist it. For example, when Emma says that "it is always incomprehensible to a man that a woman should ever refuse an offer of marriage. A man always imagines a woman to be ready for anybody who asks her" (E, p. 64), she argues that men as a sex are clueless: they do not consider women as having their own preferences and making their own choices. Clueless people tend to obsess over status distinctions: in the African American tale "Malitis," a slaveowner, heavily invested in the caste difference between him and his slaves, has difficulty understanding his slaves as strategic actors and is thereby easily tricked. Cluelessness, the absence of strategic thinking, has particular characteristics and is not just generic foolishness.

Austen explores several explanations for cluelessness. For example, Austen's clueless people focus on numbers, visual detail, decontextualized literal meaning, and social status. These traits are commonly shared by people on the autistic spectrum; thus Austen suggests an explanation for cluelessness based on individual personality traits. Another of Austen's explanations for cluelessness is that not having to take another person's perspective is a mark of social superiority over that person. Thus a superior remains clueless about an inferior to sustain the status difference, even though this prevents him from realizing how the inferior is manipulating him. Austen's explanations for cluelessness apply to real-world situations, such as U.S. military actions in Vietnam and Iraq.

In this book, no previous familiarity with game theory is presumed. In the next chapter, I explain game theory from the ground up; game theory can be applied to complicated situations, but its basic ideas are not much more than common sense. I start with the concepts of choice and preferences. I discuss strategic thinking as a combination of several skills, including placing yourself in the mind of others, inferring others' motivations, and devising creative manipulations. To illustrate game theory's usefulness, I use a simple game-theoretic model to show how Beatrice and Benedick in Shakespeare's *Much Ado About Nothing*, Richard and Harrison in Richard Wright's *Black Boy*, and people revolting against an oppressive regime all face the same situation. Game theory has been criticized as capitalist ideology in its purest form— acontextual, technocratic, and a justification for selfishness. But Austen makes us rethink these criticisms, for example, in her argument that a woman should be able to choose for herself regardless of whether others consider her selfish. I conclude the chapter by looking at previous work trying to bring game theory, as well as related concepts such as "theory of mind," together with the study of literature.

Before immersing into Austen, in chapter 3 I discuss the strategic wisdom of African American folktales, such as the well-known "Tar Baby" tale. The tale of Flossie and the Fox, in which the little girl Flossie deters Fox's attack by telling Fox that she does not know that he is a fox, is an elegant analysis of power and resistance, which I also represent mathematically in chapter 4. These folktales teach how inferiors can exploit the cluelessness of status-obsessed superiors, a strategy that can come in handy. In their 1963 Birmingham, Alabama, campaign, civil rights strategists counted on the notoriously racist Commissioner of Public Safety "Bull" Connor to react in a newsworthy way, and sure enough, he brought out attack dogs and fire hoses.

In this book, no previous familiarity with Austen is presumed. I provide a summary of each novel in chapter 5, arguing that each is a chronicle of how a heroine learns to think strategically: for example, in *Northanger Abbey*, Catherine Morland must learn to make her own independent choices in a sequence of increasingly important situations, and in *Emma*, Emma Woodhouse learns that pride in one's strategic skills can be just another form of cluelessness. Austen theorizes how people, growing from childhood into adult independence, learn strategic thinking.

Next I trace the detailed connections between Austen's novels and game theory, taking the six novels together. This is the analytic core of the book, chapters 6 to 12. Austen prizes individual choice and condemns any attempt to deny or encumber a person's ability to choose. There is "power" in being able to choose. Austen consistently delights in how completely different feelings, such as the pain of a broken heart and the pleasure of a warm fire, can compensate for each other. This commensurability, that feelings can be reduced into a single "net" feeling, is the essential assumption behind game theory's representation of preferences as numerical "payoffs," and indeed Austen sometimes jokes that feelings can be represented numerically. A person's preferences are best revealed by her choices, as in economic theory's "revealed preference"; for example, Elizabeth Bennet estimates the strength of Mr. Darcy's love by the many disadvantages it has to overcome.

Austen's names for strategic thinking include "penetration" and "foresight," and the six novels contain more than fifty strategic manipulations specifically called "schemes." For Austen, "calculation" is not the least bit technocratic or mechanistic. Austen makes fun of the strategically sophomoric; characters like Mrs. Jennings, whose manipulations are hopelessly misconceived, best illustrate (the absence of) strategic skill. The strategically skilled carefully observe others' eyes, not just because "penetration" and "foresight" are visual analogies, but because a person's eyes reveal his preferences.

Austen's commitment to game-theoretic explanation is delightfully undogmatic. She generously allows for the importance of alternative explanations, such as those based on emotion, instinct, and habit, but consistently favors explanations based on choice, preferences, and strategy. Austen's heroines make good choices even when overpowered by emotion. Even blushing, which seems to be a completely emotional response, is regarded at least partly a matter of choice. Austen acknowledges the influence of instincts and habits, but dislikes them: instinctive actions turn out badly, and habits, such as Fanny Price's submissiveness or Willoughby's idleness, are usually painful or ruinous. Twice Austen explicitly compares an explanation based on people's habits with an explanation based on their preferences and concludes that preferences are more important. Austen allows that people often follow rules or principles instead of choosing consciously, but observes that adopting a rule is itself a matter of choice.

Austen acknowledges the importance of social factors such as envy, duty, pride, and honor but in general condemns them; Austen's heroines succeed not because of social factors but in spite of them. For example, when Fanny Price receives Henry Crawford's proposal, her family members invoke social distinction, conformity, duty, and gratitude to pressure her to accept, but Fanny heroically makes her own decision based on what she herself wants. Even if social factors affect you, Austen maintains that they should affect only your behavior and not your thought processes, which must remain independent. Even under the most severe social constraints, a person can strategically maneuver; in fact, constraints make you learn strategic thinking more quickly.

Austen also takes care to distinguish strategic thinking from concepts possibly confused with it. Strategic thinking is not the same as selfishness: Fanny Dashwood is both selfish and a strategic blunderer, for example. Strategic thinking is not the same as moralizing about what one "should" do: Mary Bennet quotes maxims of proper conduct but is useless strategically. Strategic thinking is not the same as having economistic values such as frugality and thrift: Mrs. Norris exemplifies both economizing and strategic stupidity. Strategic thinking is not the same thing as being good at artificially constructed games such as card games: Henry Crawford likes to win card games but in real life cannot choose between Fanny Price and the married Maria Rushworth and fails disastrously.

In terms of results, Austen generates multiple insights not yet approached by modern game theory. In addition to analyzing cluelessness, she makes advances in four areas. First, Austen argues that strategic partnership, two people joining together to strategically manipulate a third person, is the surest foundation for friendship and marriage. Each of her couples comes together by jointly manipulating or monitoring a third person, for example a parent about to embarrass herself. Explaining to your partner your motivations and choices, strategizing in retrospect, is for Austen the height of intimacy. Second, Austen considers an individual as being composed of multiple selves, which negotiate with each other in a great variety of ways, not necessarily in a "chain of command." Just as a person anticipates other people's actions, a person can anticipate her own actions and biases; a person's self-management strategy depends on her goals. Third, Austen considers how preferences change, for example through gratitude or when an action takes on a new social connotation (for example, when rejected by a suitor, you are eager to marry another to "take revenge"). Fourth, Austen argues that constancy, maintaining one's love for another, is not passive waiting but is rather an active, strategic process which requires understanding the other's mind and motivations.

Austen even comprehensively considers the disadvantages of strategic thinking. Strategic thinking takes mental effort, gives you a more complicated moral life, allows you to better create excuses for others' misdeeds, and enlarges the scope of regret. People do not confide in you because they think you have already figured everything out; strategic skill is not charming or a sign of sincerity. Contemplating the machinations of others can be painful, and sometimes it is better to plunge ahead and not worry about how people will respond. Finally, being good at strategic thinking risks solipsism: you see strategicness where none exists, and pride in your own ability makes you think that others are perfectly knowable.

My claim is that Austen consciously intended to theorize strategic thinking in her novels; the occupation with strategic thinking is Austen's and not just mine. I do not present direct evidence for this claim (such as a letter from Austen laying bare her objectives) but a preponderance of indirect evidence. The connections between Austen's writings and game theory are just too numerous and close. Almost always when a child appears in her novels, for example, the child is either a student of strategic thinking (a three-year-old who learns to continue crying because she gets attention and sweets) or a pawn in someone else's strategic action (Emma carries her eight-month-old niece in her arms to charm away any residual ill-feeling after an argument with Mr. Knightley). After Henry Crawford's proposal to Fanny Price, Austen includes no fewer than seven examples of "reference dependence," in which the desirability of an outcome depends on the status quo to which it is compared. It is difficult to explain this repetition as a coincidence or unconscious tendency, and the remaining conclusion is that Austen explicitly intended to explore the phenomenon.

Perhaps Austen's most extensive contribution to game theory is her analysis of cluelessness. Austen gives five explanations for cluelessness, the conspicuous absence of strategic thinking. First, Austen suggests that cluelessness can result from a lack of natural ability: her clueless people have several personality traits (a fixation with numeracy, visual detail, literality, and social status) often associated with autistic spectrum disorders. Second, if you don't know much about another person, it is difficult to put yourself into his mind; thus cluelessness can result from social distance, for example between man and woman, married and unmarried, or young and old. Third, cluelessness can result from excessive self-reference, for example thinking that if you do not like something, no one else does either. Fourth, cluelessness can result from status differences: superiors are not supposed to enter into the minds of inferiors, and this is in fact a mark or privilege of higher status. Fifth, sometimes presuming to know another's mind actually works: if you can make another person desire you, for example, then his prior motivations truly don't matter. Finally, I apply these explanations to the decisive blunders of superiors in Austen's novels.

I then consider cluelessness in real-world examples and discuss five more explanations, which build upon Austen's. First, cluelessness can simply result from mental laziness. Second, entering another's mind can involve imagining oneself in that person's body, walking in his shoes, and seeing through his eyes; because of racial or status differences, a person who regards himself superior finds this physical embodiment repulsive.

Third, because social status simplifies and literalizes complicated social situations, people who are not good at strategic thinking invest more in status and prefer social environments, such as hierarchies, which define interactions in terms of status. Fourth, in certain situations cluelessness can improve your bargaining position; by not thinking about what another person will do, you can commit yourself to not reacting. Fifth, even though strategic thinking is not the same thing as empathy (understanding another's goals is not the same thing as sympathizing with them), one might lead to the other; a slaveowner for example might be easily tricked by his slaves, but if he took their point of view well enough to think about them strategically, he might not believe in slavery anymore. Finally, I apply these explanations to the disastrous U.S. attack on Fallujah in April 2004.

Why do people do what they do? This question is too interesting to be confined to either novels or mathematical models, the humanities or the social sciences, the past or the present. I hope this book shows that it is not at all surprising that Austen, a person intensely interested in human behavior, would help create game theory.

Game Theory in Context

GAME THEORY considers interactions among two or more people and is built upon rational choice theory, which looks at the choice of a single individual. I use a simple example from Austen's *Mansfield Park* to illustrate first rational choice theory and then game theory. Since strategic thinking is the central concept of game theory, I discuss it in some detail. I then use examples from William Shakespeare's *Much Ado About Nothing* and Richard Wright's *Black Boy* to illustrate how game theory can be useful, for example, in understanding popular revolt against a regime.

The growth of game theory and rational choice theory in the social sciences has met with substantial criticism. I next discuss how recognizing Austen as a game theorist puts these criticisms in a different light. For example, some critics argue that rational choice theory glorifies selfishness and asociality, which must be held in check by social norms. But for Austen, insisting upon the right to choose according to one's own preferences (over whom to marry, for example) is not selfish but subversive. Austen's heroines are already ensnared in social obligations and expectations; more social norms are exactly what they do not need.

Game theory and literature have had previous interactions; for example, a few game theorists have analyzed literary works. Again, my stronger claim is that Austen herself is a game theorist, who in her novels explores decision-making and strategic thinking systematically and theoretically. Finally, I consider how humanists have used ideas about rationality and cognition to analyze literature, and how understanding Austen as a game theorist relates to these discussions.

RATIONAL CHOICE THEORY

In Austen's *Mansfield Park*, five-year-old Betsey and her fourteen-year-old sister Susan are fighting because Betsey has taken possession of Susan's knife, a present from their departed sister Mary. Their older sister Fanny decides to buy a new knife for Betsey so that she will voluntarily give up Susan's knife. Betsey takes the new knife and household peace is restored.

Consider the situation before Fanny intervenes. Betsey chooses either to keep Susan's knife or give it up. Rational choice theory represents this with the following diagram:

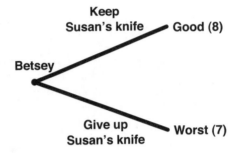

Figure 1. Betsey's decision.

Here Betsey's two choices are represented by two lines or "branches" coming from a dot or "node." This node has Betsey's name on it because Betsey is choosing. One branch is labeled "Keep Susan's knife" and the other is "Give up Susan's knife." If she chooses the branch of keeping Susan's knife, she considers the outcome good, and we write "Good." But if she chooses the branch of giving it up, she has nothing, which she considers "Worst." This "tree" is a simple way to represent Betsey's decision.

One convenient way to represent Betsey's preferences is to use numbers. For Good, we can write the number 8, and for Worst we can write the number 7, the idea being that Betsey prefers higher numbers. The numbers are called "payoffs" or "utilities."

Once Fanny offers Betsey a new knife, however, the situation changes. Betsey still chooses whether to keep or give up Susan's knife, but her preferences are different. Keeping Susan's knife is still good, but now giving it up means getting the new knife, which she likes best. So now we have a different tree. Here we write the payoff 9 for Best.

Figure 2. Betsey's decision after Fanny buys a new knife for her.

Almost all of rational choice theory boils down to trees like these. The core model of rational choice theory is payoff maximization: a person has numerical payoffs for each alternative and chooses the one with the highest payoff.

Assigning payoff numbers to outcomes might seem artificial and crude, but this is merely a convenient way to notate a person's ranking from best to worst. The implications of payoff maximization are best illustrated in examples. Say Violet decides how many children to give birth to and raise during her life. Realistically, her possible choices are 0, 1, 2, 3, 4, or 5 children. She decides that she would be most content without children, that is, choosing 0. Upon a medical examination, however, she is informed by her doctor that she can have at most one child, and so her set of alternatives is now 0 and 1. Knowing that she can have at most one child, she decides that she should have one child and chooses 1. Violet's choice is understandable, but violates payoff maximization (assuming that there are no "ties" and that her payoffs themselves do not change when she sees the doctor). Choosing 1 from the two alternatives 0 and 1 means that 1 has a higher payoff than 0, which means that 0 could not have had the highest payoff in the original situation.

For another example, say that Walter spends his money only on coffee, beer, and cigarettes, and one day his income doubles. Once his income doubles, he might consume more of all three items, he might consume less of some items and more of others, or his consumption of all items might remain unchanged. But it is not possible for his consumption of all items to decrease, if we assume payoff maximization, because he could have done the same thing before his income doubled, and he didn't. If his consumption of all items decreases, payoff maximization is violated.

One might think that a crucial issue for rational choice theory is what "rationality" means, but it is not. The term "rational" is often associated with instrumental, calculated, calm, deliberate, knowledge-able, individualistic action and is often contrasted with impetuousness, emotionality, ignorance, ideological bias, sentimentality, and social-mindedness. Rational choice theory at its core, however, is about none of these things. According to rational choice theory, a person makes a "rational choice" if it can be described by payoff maximization; for example if Walter consumes less of all items when his income doubles, he is not making a rational choice. Payoff maximization does not translate directly into any intuitive or colloquial conception of "rationality"; an altruistic person is no more or less likely to violate payoff maximization than a selfish person, for example. Rational choice theory also does not care about what the alternatives actually are; all that matters is that a person chooses among them in a way consistent with the model. A person with one hundred dollars might choose between getting a fancy haircut, donating the money anonymously to the local food bank, giving the money to his itinerant brother, or buying a handgun and shooting himself. A vain person, a generous person, a family-minded person, and a suicidal person can all be described by payoff maximization.

Payoff maximization is a purposefully crude way to describe how people make choices, especially compared to psychological studies of decision-making, for example. If Walter consumes less of everything after his income doubles, payoff maximization is violated, regardless of whether Walter makes this decision in a calm, thoughtful, instrumental, individualistic, or calculating manner, or whether he makes this decision out of habit, intuition, superstition, rules of thumb, in a fit of anger, or because of social pressures. Violet's decision about how many children to have can be made with a mixture of prudence and impulsiveness, and might involve a messy mixture of financial constraints, lifestyle changes, emotions such as guilt and joy, celebration of her newly valued fertility, concerns for her family, and her own identity as a woman and mother. For the outside observer, explaining a person's fertility choices is not necessarily easy, and for that matter Violet herself could spend years retrospectively trying to understand this single decision.

However, even with a crude model of how people make choices, things can get complicated enough, especially when two or more people are involved.

GAME THEORY

In our *Mansfield Park* example, Betsey is not the only person making a choice. Fanny also chooses whether to buy a new knife or not, and anticipates what Betsey will do in response. To consider more than one person, we need game theory.

Our earlier two trees describe two situations: the first (figure 1) describes Betsey's decision if Fanny does nothing, and the second (figure 2) describes Betsey's decision if Fanny buys a new knife. But which of these two situations takes place is up to Fanny. Hence we build these two trees into a larger tree, as shown in figure 3.

Here Fanny makes the first decision, and her node is the one at the left. She can either choose to do nothing or buy a new knife. If Fanny does nothing, then we have the first tree we derived before which represents Betsey's situation before Fanny intervenes. If Fanny buys a new knife, then we have the second tree we derived before. For clarity, Betsey's actions and branches are in boldface, and Fanny's are in regular type.

The only thing remaining is to include Fanny's preferences. The status quo in which Betsey keeps Susan's knife is Bad (we give this payoff 2),

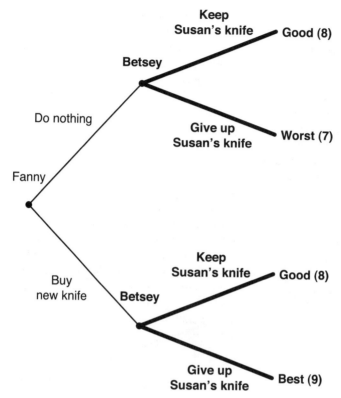

Figure 3. The two trees make a larger tree.

but the Worst thing for Fanny is if Betsey keeps Susan's knife even after Fanny buys a new one (payoff 1), since Fanny's purchase would be for nought. If Betsey gives up Susan's knife once offered a new knife, that is a Good outcome (payoff 3). The Best outcome for Fanny is if Betsey gives up Susan's knife without Fanny having to buy anything (payoff 4). Again, we use boldface to indicate Betsey's preferences and regular type for Fanny's, and we have the tree shown in figure 4.

This tree, which is called a "game tree" or "extensive form game," captures the fact that once Fanny decides whether to buy a new knife or not, it is Betsey's decision whether to keep Susan's knife or give it up. It shows how Fanny, before making her decision, must think about how Betsey will choose in response.

Writing down and analyzing diagrams like these is what game theory does. More complicated situations—for example, if people can play a series of moves and countermoves—can be represented in the same way, just with more elaborate trees. Computer programs that play chess, for

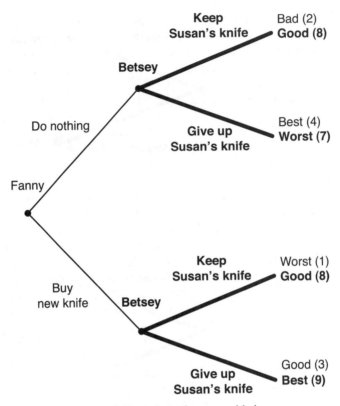

Figure 4. The larger tree with Fanny's preferences added.

example, construct very large ones. Game theory is pretty light in what it "imposes": if you describe a situation like Fanny's manipulation of Betsey, you pretty much have to specify at a minimum which people are involved, what their possible choices are, and how they feel about the possible outcomes. In this sense, a game tree is a kind of minimal notation, like musical notation, which specifies note pitch and duration, not phrasing or various kinds of expressiveness.

The tree shows how Fanny's choice "interacts" with Betsey's choice: Fanny, when making her choice, must consider how Betsey will choose. Game theory focuses on this interaction and thus does not usually consider more psychologically realistic or subtle models of choice. For example, when Fanny considers whether to buy a new knife for Betsey, she need not think much about whether Betsey is jealous of the deceased Mary's love for Susan and therefore covets the knife that Mary gave to Susan, whether Betsey likes the knife because it symbolizes power and autonomy, or whether Betsey simply likes shiny metal objects. For the sake of the manipulation, Fanny needs only a crude model: Betsey prefers a new knife over Susan's knife over no knife. For us to understand why Fanny buys a new knife, similarly we do not have to peer into Fanny's soul; we just have to understand that Fanny is willing to pay for a new knife if Betsey gives up Susan's knife. Again, individual choices can be quite complicated, but we model them crudely in order to focus on their interaction, how each person's choice depends on the choices of others.

STRATEGIC THINKING

To manipulate Betsey, Fanny must think about how Betsey will choose when offered a new knife. Fanny must engage in strategic thinking, which actually involves several related skills.

First, Fanny must realize that the person doing the choosing is Betsey, not herself. This realization might be considered obvious, but it cannot be taken for granted. Understanding that another person's mind is different from yours requires a "theory of mind," a skill that most people acquire in early childhood. For example, in the "false belief" test, a child is shown a scene "acted out" by two dolls, Sally and Anne. Sally puts a marble in her basket, and then leaves. While Sally is gone, Anne takes the marble and hides it in her box. Sally returns and the experimenter asks the child where Sally will look for the marble. Children of ages four and up typically point to the basket, while younger children point to the box (Wimmer and Perner 1983; Baron-Cohen, Leslie, and Frith 1985; Bloom and German 2000). Very young children are not yet skilled at knowing how another person's beliefs can be different from their own.

A person's theory of mind depends on her environment as well as her age. For example, it helps to be exposed to minds that are different from your own; among three- to five-year-olds, children who have nontwin siblings and especially children with at least one opposite-sex sibling are generally better at false belief tests than only children and twins with no other siblings (Cassidy, Fineberg, Brown, and Perkins 2005). Theory of mind can be understood as culturally specific. For example, in Europe and the United States in modern times, people often explain action by mental states as opposed to spirits, gods, or life forces. In contrast, Samoan children do not offer "I did not mean to" as an excuse for bad behavior (Lillard 1998). Some languages, such as Turkish, have a specific word for believing something that is not true, in contrast to English, for example, which uses "think" both for believing something true and believing something untrue. Using this specific word when asking children improves their performance in false belief tests (Shatz, Diesendruck, Martinez-Beck, and Akar 2003). One might speculate that believers in a single all-powerful god might be worse at understanding others' mental states than people who must track several interacting gods (Chrissochoidis and Huck 2011). Theory of mind is evidently confined to humans. For example, chimpanzees do not appear to have a theory of mind; a monkey will make a gesture asking a human experimenter for food even if the experimenter is wearing an obvious blindfold (Povinelli and Vonk 2003).

Some argue that people on the autistic spectrum have weak theory of mind skills, and even that this weakness is an essential aspect of autism (Baron-Cohen 1997). On the other hand, there is evidence that children on the autistic spectrum do fine when false-belief tests are presented in visual terms. In one experiment, children are asked to draw a red apple with a green pen, and are then asked what color apple did they intend to draw, as well as what color apple would a person who just walked in think that they drew. Children on the autistic spectrum report that they themselves intended to draw a red apple but that another person who sees only the drawing would think that they drew a green apple (Peterson 2002; see also Gernsbacher and Frymiare 2005). The animal behavior specialist Temple Grandin states that her theory of mind operates visually (Grandin 2008). Keen attention to visual detail is common among people on the autistic spectrum, and Grandin finds that this helps her understand how animals perceive their surroundings (Grandin and Johnson 2004).

A related characteristic of people on the autistic spectrum is a strong orientation toward literality, "relaying factual information or phrases memorized from TV shows without responding to what their listener is saying or doing....[T]hey may hear the saying 'Don't let the cat out of the bag' and search for a cat and bag" (Baker 2001, p. xi). Similarly, understanding sarcasm requires knowing how a person's intended meaning differs from her literal words; people with impaired theory of mind have difficulty detecting sarcasm (Shany-Ur, Poorzand, Grossman, Growdon, Jang, Ketelle, Miller, and Rankin 2012).

Once Fanny's theory of mind skills are established and she understands that Betsey is making her own choice, Fanny must think about how Betsey will make that choice. Fanny must consider what Betsey wants, in other words Betsey's preferences. Fanny correctly guesses that although Betsey prefers Susan's knife to nothing at all, Betsey prefers the new knife to Susan's knife. Just as one must learn that other people's minds are different from your own, one must learn that other people's preferences can differ from your own. This also is obvious, but mistakes are common, for example when you enjoy a book so much that you buy it for a friend without taking a moment to consider whether she will actually care for it. Understanding another person's preferences can be surprisingly difficult; for example, if you detest cigarette smoking or deer hunting or watching soap operas yourself, you might have real difficulty thinking that a friend enjoys these activities.

One technique to better understand another person's preferences and choices is "perspective taking," in which you consciously try to place yourself in the mind of another. Physical or visual analogies are often used, as in "put yourself in her place" or "see it from his point of view." To help its younger designers better understand the preferences of older car buyers, the Nissan Motor Company developed an "aging suit," which includes cloudy goggles, a weighted belt, and constraining elastic bands; by wearing it, one thereby adopts an older person's diminished vision, heavy midsection, and limited flexibility (Neil 2008). Taking another's perspective helps one empathize, but perspective taking and empathy are not quite the same. Empathy is more about sharing feelings. For example, you can put yourself in the mind of your military adversary to understand his objectives and predict his choices, without feeling the pain of your adversary's casualties.

To infer the motivations of others, you can observe their actions and statements, as well as their facial expressions and body language. Even when you know a person well it is not always easy to figure out her preferences; for example, when your mother says on the telephone that she will not be disappointed if you do not come home for the holidays, it can take substantial effort—listening to her tone of voice and interpreting her side remarks—to figure out, even imperfectly, how she really feels. In Korean, this skill is called *nunchi*. A person with good *nunchi* can understand another's desires when they are not expressed explicitly, can size up a social situation quickly, and can use this skill to get ahead; for example, by using her *nunchi* to understand how she can help the wife of her husband's boss, a woman might help her husband get promoted (Shim, Kim, and Martin 2008, p. 94).

The literal meaning of *nunchi* is "eye-reading," and indeed much research on understanding the minds of others is about how people see each other's eyes. In one experiment, men are worse than women at identifying people's mental states (for example, if a person is happy, sad, angry, or afraid) from pictures of their eyes, and people on the autistic spectrum are worse than non-autistics; autism has been interpreted as an extreme form of the male brain (Baron-Cohen, Jolliffe, Mortimore, and Robertson 1997; Baron-Cohen 2002). Also, by looking at someone's eyes, you can see what they are seeing. When I talk to you and we make eye contact, I know that you are paying attention; also, since you see me observing you, you know that I know that you are paying attention, and so forth (Chwe 2001). Compared to eighty other primate species, humans are the only species that have unpigmented white as opposed to pigmented brown sclera (whites of the eyes), and humans also have the largest exposed sclera area; the interpretation is that these adaptations enabled humans to better see each other's gaze direction (Kobayashi and Kohshima 2001). Great apes follow a human's gaze more by looking at the direction of the human's head than his eyes (for example, they follow a human's head direction even when the human's eyes are closed), while even one-year-old human infants follow eyes (Tomasello, Hare, Lehmann, and Call 2007).

Being able to understand the mind of another does not mean that you always do so. In one experiment, the "participant" is asked to put a roll of tape into a paper bag while the "director" is behind a large cardboard wall and obviously cannot see what the participant is doing. The wall is then removed and the paper bag is placed, along with other objects, including a cassette tape, between the director and participant. The director then asks the participant to "move the tape." Since the participant has a theory of mind, she should understand that the director does not know that there is a roll of tape in the paper bag and thus must mean the cassette tape. But most participants reach for or grab the bag before correcting themselves; almost all look at the bag. So even though the participant knows that the director could not possibly mean the roll of tape in the bag, the participant still cannot help looking at it and even reaching toward it. The interpretation is that theory of mind is like an espresso machine you are given as a present, which you leave "in the box" until you actually need it (Keysar, Lin, and Barr 2003).

A person's strategic thinking might thus depend not only on her ability and training but also on the kind of situation in which she "takes it out of the box." For me, a cocktail party might hardly be the place for goal-directed strategic reasoning, while for you it might be ideal. I might be terrible at strategic thinking in open-ended situations like cocktail parties, which have no explicit "rules" for behavior, but be very good at it in explicitly defined situations like chess, while you might be the opposite. In one experiment, people on the autistic spectrum are just as good as a control group, even slightly better, at responding correctly to "move the tape" requests, but when asked to retell a story, use fewer terms referring to characters' mental states (Beeger, Malle, Nieuwland, and Keysar 2010). Perhaps people on the autistic spectrum do not take their theory of mind skills "out of the box" when recounting a story but employ them easily when concrete action is necessary (see also Sally and Hill 2006). In another experiment, Chinese students are better than non-Asian U.S. students at responding correctly to "move the tape" requests; the interpretation is not that Chinese students are more able, but that they are more accustomed to using their theory of mind, since they participate in a culture that emphasizes "interdependent selves" (Wu and Keysar 2007).

To think strategically, one must also take into account how others think strategically. In Austen's *Sense and Sensibility,* Mrs. Dashwood, thinking that her daughter Marianne has cleverly arranged to be alone at home to receive her suitor Willoughby, therefore excuses Marianne from the family's social visit to Lady Middleton. If Mrs. Dashwood had thought Marianne strategically unsophisticated, she would have asked Marianne to come along. In general, you might take an action thinking that another person will take an action because a third person will take an action to prevent a fourth person from taking an action, and so forth. Thus strategic thinking involves both estimating the strategicness of others and also the more computational process (as in playing chess) of sorting through sequences of actions and reactions. When Renfroe (2009, p. 157) prepares for an argument with her husband, "I think of what it is that I need to say to him and three possible responses. Then I think of three things I could say to each of those responses. Within minutes I am already twenty-seven moves into the conversation, and he doesn't even have a clue we need to talk yet."

Perhaps the most advanced skill involved in strategic thinking is coming up with manipulations and plans, creating situations in which people act in such a way as to produce the desired outcome. Fanny's idea of buying a new knife for Betsey is not especially creative, but still not everyone would have thought of it. Some strategic manipulations are strikingly clever. For example, Rabbi Harvey, carrying his friend's precious candlestick, is held up by a thief. Rabbi Harvey asks the thief to shoot bullet holes in his jacket and hat so that he can prove to his friend that he was robbed; by asking the thief repeatedly, Rabbi Harvey depletes all of the thief's ammunition (Sheinkin 2008). For another example, when a woman who wants to keep all of her late husband's inheritance refuses to acknowledge that a young boy is her son, Ali (son of Abu-Talib) orders her to marry the boy, and thus she admits the truth (Khawam 1980, p. 143). Coming up with effective plans involves creativity and ingenuity and is not so easy to teach. The best way seems to be the case study method, discussing those manipulations that are particularly unexpected and brilliant.

How Game Theory Is Useful

Game theory has been used in many different ways in the social sciences, to understand, for example, how people cooperate, how workers, managers, and legislators bargain, why nations fight wars, why people join social movements, how firms compete with each other, and why Super Bowl television commercials are so expensive (for a general introduction, see for example Dixit and Nalebuff 2008; on Super Bowl advertising and social movements, see Chwe 2001). Here I consider examples of what game theory calls "coordination problems." Game theory is most useful in drawing connections among seemingly quite different situations: here I consider Beatrice and Benedick in William Shakespeare's *Much Ado About Nothing*, Richard and Harrison in Richard Wright's autobiographical *Black Boy*, and citizens revolting against a regime.

In *Much Ado About Nothing* (Shakespeare 1600 [2004]), Beatrice and Benedick greet each other with insults and disdain. However, Beatrice's family (her uncle Leonato, her cousin Hero, and Hero's attendant, Ursula) and Benedick's friends (Don Pedro and Claudio) manipulate them into believing that each has a secret love for the other, and thus each falls for the other, making the fabrication true. They need outside help because of their pride; Beatrice explains to Don Pedro that she insults Benedick "[s]o I would not he should do me, my lord, lest I should prove the mother of fools" (p. 231). When they realize that they have been manipulated, their love momentarily falters but is saved by the evidence of love poems each has secretly written, stolen from their pockets by Hero and Claudio.

Beatrice and Benedick each must choose whether to love or not, and neither knows the other's choice before making his or her own (in comparison, Betsey chooses knowing whether Fanny bought a new knife). There are four possible outcomes: both love, only Beatrice loves, only Benedick loves, and neither loves. We can write these outcomes in a table. For clarity, we write Benedick's action in boldface.

TABLE 1

	Benedick loves	**Benedick doesn't**
Beatrice loves	"Benedick, love on; I will requite thee."	"I should prove the mother of fools."
Beatrice doesn't	"Stand I condemned for pride and scorn so much?"	"No, Uncle, I'll none."

Each of the four outcomes is represented by a quote from Beatrice expressing her opinion of that outcome (pp. 237, 231, 237, 229). If both love (the upper-left entry in the table), then Beatrice joyfully returns her love: "I will requite thee" (p. 237). If only Beatrice loves (the upper-right entry), then Beatrice feels foolish and embarrassed. If only Benedick loves (the lower-left entry), Beatrice is happy but feels bad for being so scornful. If neither loves (the lower-right entry), Beatrice tells her uncle Leonato that she is content marrying no man: "No, Uncle, I'll none.

Adam's sons are my brethren, and truly I hold it a sin to match in my kindred" (p. 229).

We can abbreviate the feelings behind Beatrice's quotes in the following way:

TABLE 2

	Benedick loves	Benedick doesn't
Beatrice loves	Best	Worst
Beatrice doesn't	OK	Bad

The Best thing for Beatrice is if they both love and the Worst thing is to love without being loved. Being loved but not returning it is OK, and neither loving is Bad but better than being a fool. Note that if Benedick does not love (the right column), then Beatrice does not want to love either. If Benedick does love (the left column), then Beatrice wants to love.

Benedick's feelings are similar. The Best thing for him also is if both love, the Worst is if he loves without being loved back, and neither loving is Bad. So his feelings look like the following (again, we use boldface to distinguish his feelings from Beatrice's):

TABLE 3

	Benedick loves	Benedick doesn't
Beatrice loves	**Best**	**OK**
Beatrice doesn't	**Worst**	**Bad**

The difference between Beatrice's and Benedick's tables is that for Beatrice, the worst thing is if her love is unrequited (the upper right outcome) and for Benedick, the worst thing is if his love is unrequited (the lower left outcome).

For compact exposition, we merge these two tables (table 2 and table 3) together into a single table. Here, in each of the four outcomes, we first write Beatrice's feelings and then Benedick's.

TABLE 4

	Benedick loves	Benedick doesn't
Beatrice loves	Best, **Best**	Worst, **OK**
Beatrice doesn't	OK, **Worst**	Bad, **Bad**

Again, the two agree on what is best (mutual love) and what is bad (mutual indifference). What is worst for Beatrice (loving Benedick foolishly) is second-best for Benedick, and what is worst for Benedick is second-best for Beatrice.

This table, called a "strategic form game," distills the Beatrice-Benedick situation to its essential elements. It might seem slightly complicated at first, but it cannot be made any simpler. Love does not come upon them like a fever or euphoria; each consciously chooses to love. Each knows that choosing to love risks foolishness. Each is painfully aware that all four outcomes are possible, and that by trying for the best, one risks the worst. One cannot simply say that each desires the other; it is essential to the situation that each person wants to love only if the other does also. One also cannot simply say that Beatrice and Benedick "find love" with the help of their friends and thus collectively move from bad to best; they are both independent individuals who make independent choices, and their love almost unravels once they are informed of their friends' manipulation.

In *Black Boy*, Richard Wright was at his job washing eyeglasses when Mr. Olin, his white foreman, approached to tell him that Harrison, a worker at a rival optical house, had a grudge against him (Wright 1945 [1993], pp. 235–37). " 'Well, you better watch that nigger Harrison,' Mr. Olin said in a low, confidential tone. 'A little while ago I went down to get a Coca-Cola and Harrison was waiting for you at the door with a knife. . . . Said he was going to get you.' . . . 'I've got to see that boy and talk to him,' I said, thinking out loud. 'No, you'd better not,' Mr. Olin said. 'You'd better let some of us white boys talk to him.' "

Richard seeks out Harrison anyhow. " 'Say, Harrison, what's this all about?' I asked, standing cautiously four feet from him. . . . 'I haven't done anything to you,' I said. 'And I ain't got nothing against you,' he mumbled, still watchful. . . . 'But Mr. Olin said that you came over to the factory this morning, looking for me with a knife.' 'Aw, naw,' he said, more at ease now. 'I ain't been in your factory all day.' . . . 'But why would Mr. Olin tell me things like that?' I asked. Harrison dropped his head; he laid his sandwich aside. 'I . . . I . . .' he stammered and pulled from his pocket a long gleaming knife; it was already open. 'I was just waiting to see what you was going to do to me . . .' I leaned weakly against a wall, feeling sick, my eyes upon the sharp steel blade of the knife. 'You were going to cut me?' I asked. 'If you had cut me, I was going to cut you first,' he said."

Harrison is not a fool for carrying the knife; as he says, if you think that the other will bring a knife, you would want to bring one also. In this situation, Richard and Harrison each choose either to bring a knife or not. We can make a table as before. Here Richard's feelings are in regular type and Harrison's are in boldface.

TABLE 5

	Harrison doesn't	Harrison brings knife
Richard doesn't	Best, **Best**	Worst, OK
Richard brings knife	OK, **Worst**	Bad, **Bad**

For both Richard and Harrison, the best outcome is if neither brings a knife and life goes on normally. If you bring a knife and the other does not, that's OK but not the best, since you are embarrassed for revealing your distrust. If both bring a knife, that is bad for both, but the worst thing is if you don't bring a knife and the other does. So if the other doesn't bring a knife, you don't want to either. But if the other brings a knife, you would be stupid not to bring one also.

Richard and Harrison vow to keep faith in each other and ignore their white bosses' provocations. But when each is offered five dollars to fight the other in a boxing match, Harrison convinces a reluctant Richard, saying that it's just exercise and they can fool the white men into thinking they are really hurting each other. However, "[w]e squared off and at once I knew that I had not thought sufficiently about what I had bargained for. . . . The white men were smoking and yelling obscenities at us. 'Crush that nigger's nuts, nigger!' . . . [B]efore I knew it, I had landed a hard right on Harrison's mouth and blood came. Harrison shot a blow to my nose. The fight was on, was on against our will. I felt trapped and ashamed. I lashed out even harder, and the harder I fought the harder Harrison fought. Our plans and promises now meant nothing. . . . The hate we felt for the men whom we had tried to cheat went into the blows we threw at each other. . . . [E]ach of us was afraid to stop and ask for time for fear of receiving a blow that would knock us out. When we were on the point of collapsing from exhaustion, they pulled us apart. I could not look at Harrison. I hated him and I hated myself" (pp. 242–43).

How were their actions "against their will"? They both had agreed to pretend, but once the other begins to fight in earnest, even accidentally, each must fight in return, making things bad for both.

The Beatrice-Benedick situation and the Richard-Harrison situation seem quite different. One is delightful and the other is sobering. One is an unexpected triumph and the other is a degrading defeat. One is about love and the other is about hate.

But when we use the tables above to distill each situation, we find that the two situations are identical. The table that describes the Beatrice-Benedick situation (table 4) and the table that describes the Richard-Harrison situation (table 5) are identical, different only in the names of the characters and the names of their actions. In both situations, the two people involved have a "good but risky" action (loving, not bringing a knife) and a "bad but safe" action (not loving, bringing a knife). The best for both people is if both take the good but risky action, but taking that action without the other doing so yields the worst possible outcome. Taking the good but risky action requires an assurance that the other will do the same (Sen 1967).

We might have discovered this similarity without all this apparatus. But the tables make it much easier. Once we have pedantically written down the tables, finding the similarity is a matter of inspection. Once we see the similarity, it becomes clear how mutual love and mutual hatred both can be created out of nothing, and in what sense exactly this creation is against their wills. It becomes clear how one person's action can be provoked by nothing more than her own expectation of the other person's action, and that once provoked, each person's action can in turn respond to the other's action, resulting in an unexpectedly good or bad outcome, a virtuous or vicious cycle. The third-party manipulators (Hero, Leonato, Ursula, Don Pedro, and Claudio, and Mr. Otis and the other white foremen) have opposite goals but the same method: influencing the expectations of each person about the other in a way that becomes self-confirming.

Our tables are, of course, abstractions; in any abstraction, something is lost, but what is gained is the possibility of finding connections among seemingly disparate things. Whether this gain is worth the loss is best decided in specific contexts. Here the connection between Beatrice-Benedick and Richard-Harrison is, I think, at least slightly unexpected.

When people revolt against an oppressive regime, we make a table like the following:

TABLE 6

	Person 2 revolts	Person 2 stays at home
Person 1 revolts	Best, **Best**	Worst, **OK**
Person 1 stays at home	OK, **Worst**	Bad, **Bad**

Here we simplify a society into just two people. If everyone revolts, then the regime is overthrown, which is best for everyone. However, if you revolt and others stay at home, then you get shot at, which is the worst outcome. If everyone stays at home, the bad status quo remains. If you stay at home and others revolt, that's OK from your point of view but not the best.

This table is identical to our previous two tables. Just as Beatrice wants to love only if Benedick does the same, and just as Richard wants to not bring a knife only if Harrison does the same, a person wants to revolt only if enough others do the same. Everyone wins if everyone takes the good but risky action, but no one wants to do so alone. Therefore, just as in the Beatrice-Benedick situation and the Richard-Harrison situation, what is essential is everyone's expectations about everyone else. Like Mr. Olin, oppressive regimes try to make people doubt that anyone else will revolt, knowing that this doubt can be self-perpetuating. Like Beatrice's and Benedick's friends, people working against a regime try to create an optimism that feeds on itself.

A wide variety of social situations can be represented by this same table. These situations are called "coordination problems" (see Chwe 2001 for example). Adopting new technology (if enough of my friends use the latest social networking website, I want to start), seeing a movie (I want to see a movie more if it is popular), finding love, sustaining nonviolence, and joining a protest are in a fundamental respect all the same and can all be analyzed in the same way. Richard Wright perhaps intended the situation of Richard and Harrison to be a parable for African American political mobilization. Game theory allows us to understand all of these related situations as parables for each other.

CRITICISMS

The growth of game theory and rational choice theory has not been universally welcomed (for example, Archer and Tritter (2000) describe it using the terms "colonization" and "imperialism"). The various points of contention have been often surveyed (for example, Friedman 1996). Austen is far above such controversies, and of course wrote before the drawing of disciplinary boundaries; in chapters 7 and 8, I discuss how she judiciously considers alternatives to game-theoretic explanations and carefully distinguishes between strategic thinking and concepts such as selfishness. Here I consider various criticisms and discuss how recognizing Austen as a game theorist puts them in a new light; for example, if game theory was developed by Austen and slave folktellers, it cannot be understood as an intellectual handmaid of more recent historical developments such as capitalism.

As mentioned before, rational choice theory at its core is the payoff maximization model. The payoff maximization model says nothing about what a person's preferences are, whether a person wants to be expressive or instrumental, altruistic or selfish, cruel or kind. The payoff maximization model is also not meant to describe the actual process by which a person makes a choice. When Sethe kills her infant daughter in Toni Morrison's *Beloved* (1987) because she does not want her to live life as a slave, whether this decision is "rational" is not a question that rational choice theory tries to answer. A grandparent who travels long distances at great cost just to dote on a grandchild (Abelson 1996) does not violate rational choice theory.

A common criticism is that game theory and rational choice theory assume "passionless, atomistic individuals," unaffected by social mores or norms, outside any social or cultural context (Archer 2000, p. 50). In the Beatrice-Benedick and Richard-Harrison situations, however, people live in a thick social milieu, complete with expectations about courtship and how black people can talk to whites, for example, and within dense networks of affection and distrust. Richard and Harrison try to create their own norm by vowing to trust each other, but it falls apart. It would be odd to say that carrying a knife, avoiding the risk of heartache, or punching back in desperation is atomistic, selfish, or self-interested as opposed to holistic, altruistic, or public-spirited. It is hard to say that our tables impose an "individualistic" logic: Beatrice and Benedick's table (table 4) does little more than notate that they both gain if they both love, but neither wants to be a fool and love without being loved back. By offering the new knife to Betsey, all Fanny wants is to restore peace in the household. Is Fanny atomistic?

As for "passionless," the Beatrice-Benedick and Richard-Harrison situations are steeped in fear, joy, anticipation, disappointment, shame, and disgust, and indeed we took these emotions into account when we wrote down which outcomes were Worst, Best, Bad, and OK in the tables. People often make the most careful and consciously important choices, such as whether to take a feverish wailing child to the emergency room, not in placidity but in uproar. Beatrice and Benedick each make a conscious decision to love, but this does not make the achievement of their love any less emotional. Emotions can be important in many ways not captured in our tables, but it cannot be said that when we write down our tables, or assume that people make conscious choices, we exclude emotion outright.

Some argue that rational choice theory legitimizes capitalism (Amadae 2003, for example), partly because it seems difficult to condemn a system in which everyone is said to be making choices. But the fact that a victim chooses to hand over his wallet at gunpoint does not make the perpetrator's actions any less criminal. Saying that slaveowners were profit maximizers and not sadists does not legitimize slavery (see, for example, Chwe 1990). The fact that a woman chooses to remain in an abusive relationship does not excuse the abuse.

For that matter, rational choice arguments can legitimize other things too. Until the 1970s, scholars explained social protest in terms of "mob mentality": "crowds were assumed to create, through suggestion and contagion, a kind of psychologically 'primitive' group mind and group feelings" (Goodwin, Jasper, and Polletta 2001, p. 2). Protests were understood as "something irrational others engaged in" (Calhoun 2001, p. 48). What changed in the 1970s was simply that scholars began to think that social protests were normal, partly because they themselves had participated. Participating in social movements such as the civil rights, lesbian and gay rights, feminist, and environmental movements became "something that 'people like us' might do. It was seen as rational in the sense of reasonable, self-aware product of choice as well as (more narrowly) strategic, interest-based, calculated in terms of efficient means to an end" (Calhoun 2001, p. 48). For another example, Walkowitz (1980, p. 9) examines the struggle of Plymouth and Southampton prostitutes in the 1870s against the Contagious Diseases Acts and concludes, "Prostitutes thus emerge as important historical actors, as women who made their own history, albeit under very restrictive conditions. They were not rootless social outcasts but poor working women trying to survive in towns that offered them few employment opportunities and that were hostile to young women living alone. Their move into prostitution was not pathological; it was in many ways a rational choice, given the limited alternatives open to them."

Critics have quite broadly characterized rational choice theory and game theory as serving political trends. Amadae (2003, p. 9) argues that rational choice theory and game theory should be understood as "a philosophic underpinning for American economic and political liberalism," and emphasizes the origins of game theory in the early U.S.–Soviet Cold War period, specifically at the RAND Corporation, established by the U.S. Air Force in 1946. Fourcade (2009, p. 128) sees the related ascendance of mathematical techniques in economics in the United States as an aspiration toward professionalism, "in the sense of a claim to objectivity, a focus on analytical capabilities, and a high degree of collective organization and regulation." Archer and Tritter (2000, p. 1) call rational choice theory "the grand theory of high modernity.... [that] has underpinned the neoliberal reforms of the public sector.... [and] the rollback of the traditional social welfare state." Taylor (2006, p. ix) calls rational choice theory "a radically reductive and dehumanizing but deeply entrenched way of thinking.... [which] has come to have enormous influence on how public policies of all kinds are made."

Rational choice theory and game theory have been employed in myriad ways and have no single inherent ideological direction. For example, most people understand violence as aberrant behavior, almost inherently irrational, resulting from emotions like aggression. This view of violence allows systemic and instrumental institutions of violence, like slavery and various instruments of the state, to elude examination. A game theoretic perspective on violence (for example, Chwe 1990), instead of letting violence float around unattached to any responsible party (as a "scourge," "cycle," or "epidemic"), focuses on why particular individuals specifically choose to hurt other people. Thinking of domestic violence as caused by emotional instability and aggression leads to remedies such as counseling, psychotherapy, and medication (see Gordon 1988 for a historical view). Thinking of the actors in domestic violence as making conscious choices leads to remedies that are often considered feminist victories: harsher criminal penalties that decrease the incentive to batter, and women's shelters that give battered women better choices. Bancroft (2002, p. 34) writes, "While a man is on an abusive rampage, verbally or physically, his mind maintains awareness of a number of questions: 'Am I doing something that other people could find out about, so it could make me look bad? Am I doing anything that could get me in legal trouble?' "

A more diverse history of game theory is emerging. Leonard (2010) considers, for example, popular interest in chess, and its extrapolation into a "science of struggle" by twenty-four-time world chess champion and mathematician Emanuel Lasker (1907). Leonard (1995, pp. 755–56) writes that "the theory of games was intended to constitute a radical departure with the Hicks-Samuelson variant of neoclassical economics. . . . [T]he break with mechanism for the analysis of *structure,* evident in the theory of games, was a shift which characterized many disciplines in the early part of the 20th century, from physics to literary criticism" (see also Leonard 1997). In the 1960s, anthropologist Claude Lévi-Strauss (1963, p. 298) and sociologist Erving Goffman (1961, 1969) were more excited about game theory than most economists. Early writers on strategic thinking include military strategist Sun-tzu (2009) and Oxford lecturer Lewis Carroll (see the survey by Dimand and Dimand 1996). Lebowitz (1988, p. 197) writes that "not only may we speculate that Marx would have been quick to explore its techniques but we can go further and suggest that Marx's analysis was inherently a 'game-theoretic' perspective."

Recognizing Austen's place in this history has the following implications (explored more fully in chapters 7 and 8). If rational choice theory and game theory are criticized for ignoring social context, Austen theorizes strategic action not in spite of but perhaps because of her unmatched sensitivity to context. Austen agrees that focusing too narrowly on contrived situations (such as a card game) can make you lose sight of the larger social context, but this is a problem only for mediocre strategists. Austen's good strategic thinkers always remain aware of the larger context and in fact use others' interest in trivial games to get them out of the way.

If rational choice theory and game theory are criticized for ignoring social norms, Austen shows how social norms, far from protecting sociality against the corrosive forces of individualism, can be the first line of oppression. For Austen, duty and decorum are often nothing but pretenses used to prevent a person from making her own choices, over whom to marry or even whether to take a walk. To control someone, call them selfish. In an environment of unbridled individualism, perhaps we should valorize social norms, but in an environment lousy with social norms like Austen's, exploring how people best use the agency they have is more urgent.

If rational choice theory and game theory are criticized for assuming that all people act like middle-class consumers, Austen elucidates the strategic wisdom of the relatively powerless. For example, Bourdieu and Wacquant (1992, p. 124) argue that "[a]ll the capacities and dispositions it [rational choice theory] liberally grants to its abstract 'actor'—the art of estimating and taking chances, the ability to anticipate through a kind of practical induction, the capacity to bet on the possible against the probable for a measured risk . . . can only be acquired under definite social and economic conditions. They are in fact always a function of one's power in, and over, the specific economy." But Austen's novels, African American slave folktales, and entire folk game theory traditions argue the opposite: the relatively powerless need strategic thinking most and learn it best. In his analysis of Polish prison life, Kaminski (2004, p. 1) writes, "Prison socializes an inmate to behave hyperrationally. . . . A clever move can shorten one's sentence, save one from a rape or a beating, keep one's spirit high, or increase one's access to resources."

If rational choice theory and game theory are criticized (England and Kilbourne 1990 and Nelson 2009) for being "masculinist" (unemotional, reductive, technical, and acontextual), Austen establishes a "women's way of strategy," analogous perhaps to "women's way of knowing" (Belenky, Clinchy, Goldberger, and Tarule 1986). Similarly, if game theory is part of the African American slave folktale tradition and other world folk traditions, it cannot be understood as an artifact of (presumably Western) high modernity.

Of course, game theory and rational choice theory have limitations, some due to our limited imaginations and some likely inherent. Any theory is worth pushing to its limits, to understand what it can and cannot explain and why. For example, Hargreaves Heap and Varoufakis (2004, pp. 3–4) write that game theory "demonstrat[es] the limits of a particular form of individualism in social science: one based *exclusively* on the model of persons as preference-satisfiers. . . . Indeed, for those who are suspicious of economic imperialism within the social sciences, game theory is, somewhat ironically, a potential ally" (here and throughout the book, emphasis is in the original). Austen is particularly thoughtful about the limits of game theory, and how strategic thinking interacts with other aspects of human behavior. For example, Austen acknowledges the importance of emotion, but intense feelings help her heroines choose better, not worse. Austen does not idealize a world in which everyone always acts strategically: she considers the disadvantages of strategic thinking (chapter 10) and even explores why people often do not think strategically (chapter 12).

A lot of game theory and rational choice theory is written in the language of mathematics and is not very accessible to nonspecialists. Mathematics is indispensable for understanding many phenomena, in the social as well as natural world; for that matter, Liu (2004) argues that the humanities should also consider itself a "technical" discipline. In any case, many essential game-theoretic insights have been expressed without mathematics (for example Schelling 1960 [1980]). Austen generates insights in language not just accessible but beloved.

GAME THEORY AND LITERATURE

Scholarly interactions between game theory and the humanities have been tentative at best (see, for example, Chwe 2009 and other essays in the collection introduced by Palumbo-Liu 2009, as well as Bender 2012 and Palumbo-Liu 2012). Daston (2004, p. 361) writes, "Rational choice theory, game theory, and other models of human conduct are frankly imperialistic in their aims. But insofar as there has been any humanistic response to them, it has been a rolling of eyes heavenward and a shrugging of shoulders about the absurdity of it all (sentiments and gestures richly reciprocated by the other side, especially the indifference)."

Game theorists occasionally employ literary examples. Morgenstern (1928, p. 98, quoted in Morgenstern 1935 [1976], pp. 173–74) considers the pursuit of Sherlock Holmes by Professor Moriarty, in which each chooses to get off the train at either Dover or Canterbury (Conan Doyle 1893 [2005]). Von Neumann and Morgenstern (1944, p. 176) call this "a paradigm of many possible conflicts in practical life." More recently, Dixit and Nalebuff (2008, p. 423) use an example from Shakespeare's *Henry V*, in which King Henry allows any soldier to leave before the battle of Agincourt, but only publicly in front of all the other soldiers (see also Dixit 2005; Watts 2002; and Watts and Smith 1989). Crawford, Costa-Gomes, and Iriberri (2010) use literary examples such as M. M. Kaye's *The Far Pavilions* in support of "level-k" models in game theory. Eliaz and Rubinstein (2011) conduct an experiment motivated by Edgar Allan Poe's (1845 [1998]) discussion of the "even and odd" game in "The Purloined Letter" (see also Deloche and Oguer 2006 and Swirski 1996).

But few have attempted to use game theory to substantially analyze literature, despite several calls to do so (including Brams 1994, reprinted in Brams 2011; De Ley 1988; Deloche and Oguer 2006; Ingrao 2001; and Swirski 1996). Brams (2002) on the Bible is one of the few book-length attempts. O'Neill (1990, 2001) uses game theory to explain why Gawain accepts a bizarre challenge in the Middle English poem *Sir Gawain and the Green Knight*, and why the knight in the poem *Lai de l'Ombre* cleverly responds to his lady's refusal of his ring by dropping the ring into his lady's reflection in a well. Chrissochoidis, Harmgart, Huck, and Müller (2010) argue that in the Richard Wagner opera, "Tannhäuser's seemingly irrational behavior is actually consistent with a strategy of redemption" (see also Harmgart, Huck, and Müller 2009 and Chrissochoidis and Huck 2011).

This book similarly uses examples from literature, like the Beatrice-Benedick and Richard-Harrison examples, to introduce game-theoretic concepts, and similarly uses game theory to analyze strategic situations in literary works. However, this book makes a stronger claim, that literary works such as Austen's novels and African American folktales are game theory, written and told with the explicit objective of theoretically analyzing strategic thinking.

This claim is similar to Livingston's (1991, p. 51) claim that "not only are the concepts and issues related to rationality and irrationality directly relevant, and indeed, essential, to enquiries concerning literature, but literature in turn has genuine cognitive value in relation to questions of human rationality and irrationality." For example, Theodore Dreiser explains a person's imitation of another in terms of biological drives or animal instinct; in *The Financier* he even invokes an animal model, the black grouper, a fish that camouflages itself to match its surroundings. However, Livingston observes that Dreiser's characters imitate in a goal-directed manner. In *Sister Carrie* (1900 [1981]), Carrie and her lover, Drouet, observe a woman walk by; Drouet remarks that she is a "fine stepper," and thus Carrie thinks to herself, "If that was so fine she must look at it more closely. Instinctively she felt a desire to imitate it. Surely she could do that too" (p. 99). Although Dreiser describes Carrie's desire to imitate as "instinctive," Livingston (1991, p. 113) points out that "Carrie's intentional attitudes and reasoning are indispensable parts of the episode: having been confronted with the proposition that a particular bit of behaviour is to be valued, she concludes that it must be observed more carefully; she asks herself whether it figures among the realm of her possible actions, [and] determines that this is indeed the case." Indeed, "the claims made by Dreiser's naturalist narrators are flatly contradicted by other aspects of the work" (p. 84). Dreiser thus interestingly contrasts with Austen, whose stated theoretical stance on human action, emphasizing preferences, choice, and strategy, is largely consistent with how her characters act.

Livingston explores irrationality by looking at the life course of Lazare Chanteau in Émile Zola's *La Joie de Vivre* (1883–84). Lazare attempts several different careers, including composing music, practicing medicine, and building a huge factory to extract valuable chemicals from seaweed, but gives up each at the slightest difficulty. Throughout Lazare thinks of each "project as a way of quickly manifesting his individual genius and singularity"; his mother is obsessed with "a public recovery of her imagined distinction and superiority" through her son (Livingston 1991, pp. 164, 176). Lazare is strategically unskilled: "he makes extremely naive judgements about the motives and capacities of others," and while negotiating with his business partner, "fail[s] to take note of the fact that he is in a strategic situation where his interests require him to formulate expectations about the possible strategic actions of the other party" (pp. 175, 168). Lazare prefers the literal, taking his mother's statements "at face value" and "attach[ing] far too much weight to what may be called erroneous 'tutelary beliefs'.... [H]e frequently assumes that the information needed is all in the hands of some single authoritative individual" (pp. 177, 175). Livingston argues that Lazare is not simply stupid; what is interesting is the particular pattern of his stupidity. In his strategic naivety and preoccupation with status and literality, we can say that Lazare joins Flossie's Fox and Austen's Mr. Collins and Sir Walter Elliot in our analysis of cluelessness, in chapters 12 and 13.

Interactions between novelists and political economists in 19th-century Britain are explored by Gallagher (2006, pp. 2, 129), who finds that "political economists and their literary antagonists had a great deal in common, which they were frequently unwilling to recognize.... So even if George Eliot never read Jevons, the similarities between his theories of the role of surfeit in economic value and her theories of the decline of aesthetic value through repetition should not surprise us, for they were conceived in overlapping intellectual circles" (see also Levy 2001). In contrast, my claim is not that Austen necessarily interacted with the social science of her time, but that her development of game theory predated what would become part of social science 150 years later, and indeed that 200 years later we are still catching up to her insights.

As for Austen's intellectual milieu, Knox-Shaw (2004, p. 5) finds that "Jane Austen is a writer of centrist views who derives in large measure from the Enlightenment, more particularly from that sceptical tradition within it that flourished in England and Scotland during the second half of the eighteenth century," which includes Adam Smith and David Hume. Knox-Shaw (2004, pp. 87–88) and Moler (1967) find Austen echoing specific passages from Adam Smith's *Theory of Moral Sentiments* (1759 [2009]), despite the argument by Rogers (2006, p. xliii) that Austen "was a novelist and we do her most serious art no service if we ask it to perform philosophic tasks in which she had little or no ascertainable interest." Knox-Shaw (2004, p. 23) argues that Austen's extreme attention to the details of social interaction is part of her empirical and scientific outlook, and "by analogy with other kinds of discourse that are empirically grounded, the 'experimental' novel does not need to offer a general theory in order to have real significance." This is sympathetic to Austen but understates her theoretical contribution.

According to Butte (2004) and Zunshine (2007), Austen was particularly innovative in analyzing how her characters think about each other. For example, when Anne Elliot sees her sister Elizabeth coldly turn away after Captain Wentworth clearly wants to be acknowledged by her, Zunshine (2007, p. 279) notes that this can be understood as involving five levels of metaknowledge: "Anne *realizes* that Wentworth *understands* that Elizabeth *pretends not to recognize* that he *wants* to be *acknowledged* as an acquaintance." Austen's novels are considered to be among the first to continuously employ free indirect style, in which it is not explicitly indicated whether a thought is a character's or the narrator's, thus creating an "apparently unmediated representation that creates the illusion of entry into the consciousness of fictional characters" (Bender 1987, p. 177; see also Finch and Bowen 1990 and Bray 2007). For Butte (2004, pp. 25–26), Austen's novels are a "sea change in the representation of consciousnesses in narratives...not only of consciousness, but also of consciousness*es*, of a newly framed intersubjectivity." Zunshine (2006) argues that literature exercises and develops the reader's theory of mind, by making the reader keep track of what each character knows about others, and that the main reason why people read fiction in general is to get this exercise. Indeed, Oatley (2011) finds evidence that people who read more fiction have better theory of mind skills.

In contrast, my claim is more specific: certain specific (not all) literary works explore and teach strategic thinking not just by making the reader follow each character's knowledge of the knowledge of others, but also by exploring each character's preferences and choices in strategic situations. Theory of mind is essential but just one part of strategic thinking. For example, Captain Wentworth asks his sister Mrs. Croft to ask the fatigued Anne Elliot to join her and Admiral Croft in their carriage, because he knows that Anne would decline his own suggestion but cannot refuse his sister's request. Here Captain Wentworth considers not just what Anne knows but also her preferences and how she will choose; coming up with the idea of asking his sister to ask Anne also requires cleverness and creativity. In other words, Austen is interested not just in interacting knowledge or intersubjectivity, but in interacting actions: how a person's choices interact with the choices of others.

Many have noticed Austen's hardheadedness on economic matters. For example, Vermeule (2010, pp. 178, 185) writes, "The more I read *Emma,* the more I realize how devastating Austen's vision of human psychology is. Her characters are locked in fierce but largely unconscious battle over a small passel of land and all the good things that flow from it.... [T]he novel is a sophisticated hydraulic system for producing a guided distribution of resources." One might say that Austen's economic "realism" is just part of her overall treatment of strategic thinking, which applies to many situations, not just economic ones. In 1949, Bill Phillips (of the "Phillips curve") built a hydraulic computer, with water flowing through tubes, valves, and tanks, to model an economy (see Leeson 2000). Since then, the standard criticism of such (in this case literally) "mechanistic" models is that they need to incorporate how people in an economy anticipate each other's actions. Also, I suggest that Austen's characters, fierce in battle, strategize consciously.

CHAPTER THREE

Folktales and Civil Rights

HERE I CONSIDER a handful of African American folktales that analyze strategic thinking (I rely on Levine 1977). In these tales, characters who do not think strategically are mocked and punished by events, while revered figures, like Brer Rabbit, skillfully anticipate others' future actions. Of course, African American folktales and their interpretations constitute an enormous literature, and trickster figures appear in many world folk traditions: "Nothing illustrates the veiled cultural resistance of subordinate groups better than what have been termed trickster tales" (Scott 1990, p. 162; see also Hynes and Doty 1993, Landay 1998, and Pelton 1980). I suggest that the African American strategic folktale tradition informed the tactics of the 1960s civil rights movement.

I start with the tale of a new slave who asks his master why the master does nothing while the slave has to work all the time (Jones 1888 [1969], p. 115, discussed in Levine 1977, p. 130). The master replies that he is working in his head, making plans and studying upon things. When the master later finds the slave resting in the field, he asks the slave why he is lazy. The slave replies that now he is working with his head, and when the master asks what kind of work his head is doing, the slave asks, "Mossa, ef you see tree pigeon duh set on dat tree limb, an you shoot an kill one er dem, how many gwine leff?" The master answers, "Any fool kin tell dat. Ob scource two gwinne leff." The slave replies, "No, Mossa, you miss. Ef you shoot an kill one er dem pigeon, de edder two boun fuh fly way, an none gwine leff." The master laughs and does not do anything to the slave when he neglects his work—"De Buckra man bleege fuh laugh, an eh yent do nuttne ter de New Nigger case eh glec eh wuk."

If the pigeons were pine cones or other inanimate objects, then the master would have been right. The master does not recognize that the pigeons are strategic actors, able to make decisions and act independently, just as humans would if shot at. The master sees the pigeons simply as objects in front of the only relevant actor, himself.

The slave's ability to think through this strategic situation is valorized, worthy indeed of the term "work," and is rewarded, with the master's laugh and more importantly his forbearance of the slave's continuing relaxation. Here the slave has anticipated strategically again. Once the master justifies his inexertion by saying that he is working in his head, the slave realizes that if he himself makes a similar justification convincingly, the master would be compelled to accept it to some degree. He thus tells the riddle anticipating the master's reaction and thereby gains materially. Here again, the master makes the same error; when he makes the initial excuse of head work, it does not even occur to him that the slave might use it in turn, because he does not recognize the slave as a strategic actor or even that slaves are biologically capable of head work.

In this brief tale there are two games: one involving the pigeons and the shooter, and one involving the slave and the master. In both games, the master makes the same mistake of not recognizing that others act strategically. By telling the riddle, the slave takes advantage of the master's mistake and at the same time demonstrates his own strategic understanding.

Masters do not think of slaves and pigeons as strategic; in this sense masters are clueless, and this tale suggests why. If you consider yourself naturally superior, a completely different kind of being, placing yourself in the mind of the inferior even for a moment is unthinkable, and indeed you might consider your not having to do so a privilege of your dominance. If you can't think of people as strategic, you completely misrecognize strategic situations involving them, and they can use your cluelessness to their advantage.

The following features are common in the folktales I examine here. Stupid people (or animals) fail to recognize that others are strategic, and fail to anticipate the actions of others. Smart people choose their actions anticipating the actions of others, get materially rewarded, and take advantage of the cluelessness of stupid people. The specific techniques of the smart, but more importantly their general aptitude in strategic thinking, are worth discussing and emulating.

In another tale (Jones 1888 [1969], p. 102, discussed in Levine 1977, p. 109), Rabbit sees the fisherman carrying fish in his wagon, and comes up with a plan to get some. Rabbit lies by the road pretending to be deathly ill, and when the fisherman stops to ask him what ails him, Rabbit says that he can't travel any farther and begs the fisherman for a ride. The fisherman agrees and places Rabbit in the wagon, where Rabbit lays down as if dead. As the fisherman proceeds down the road, Rabbit quietly throws fish one at a time into the bushes by the side of the road. When the fisherman turns off the main road, Rabbit jumps off and goes back and collects all the fish. On the way back home, Rabbit meets Fox, who asks him how he got all the fish. Rabbit tells Fox his plan, and the next day Fox tries the same trick. When he sees Fox by the side of the road, the fisherman, who of course had figured out what had happened the day before, beats Fox dead. He then takes Fox's body to his wife to show her the thief; the fisherman thinks Fox and Rabbit are the same animal. The tale ends by making clear that Rabbit knew this would happen: "Buh Rabbit, him no care so he sabe isself. Him bin know say Buh Fox gwine ketch de debble wen de Ole Man come pon topper um."

The real trickery here is not the fish-stealing, but Rabbit's framing of Fox, which does not involve any deception. Rabbit knows that when he tells Fox the plan, Fox will try it also and the fisherman will retaliate toward Fox instead. Fox's error is that he does not anticipate that the fisherman will obviously learn from the first swindle. Like the master who forgets that pigeons make their own choices once shot at, Fox forgets that the fisherman makes independent choices once tricked. The strategically sophomoric Fox is so caught up in the specific trickery of fish-stealing that he does not recognize the larger strategic setting; Rabbit takes advantage and thus gains his own innocence. Levine (1977, p. 109) interprets this tale as showing how, "unable to outwit Rabbit, his adversaries attempt to learn from him, but here too they fail." But actually Fox learns the fish-stealing technique all too well; Rabbit counts on him to do it exactly the same way he did. Fox dies not because of a failure to learn but because of a failure to see the larger strategic picture.

The "Malitis" tale is a true story from the Slave Narrative Collection of the Federal Writers' Project (Botkin 1945, pp. 4–5, discussed in Levine 1977, pp. 126–27). One master, so stingy that his slaves are almost starved, has seven hogs ready for slaughter. The day before they are to be killed, a slave boy runs and tells the master that all the hogs had died from illness. "When the master goes to where-at the hogs is laying, they's a lot of Negroes standing round looking sorrow-eyed at the wasted meat. The master asks, 'What's the illness with 'em'? 'Malitis,' they tells him, and they acts like they don't want to touch the hogs. Master says to dress them anyway for they ain't no more meat on the place. He says to keep all the meat for the slave families, but that's because he's afraid to eat it hisself account of the hogs' got malitis." What's the mysterious and fatal disease of malitis? A slave had gone to the hog pen very early that morning with a mallet, and "when he tapped Mister Hog 'tween the eyes with that mallet, 'malitis' set in mighty quick."

"Malitis" solved the problem of how the slaves could keep the meat and eat it openly (a simple theft would have required furtive consumption) by enlisting the master as a decision-maker, by motivating him to choose to transfer the meat himself. Had he thought the slaves strategic, the master would have at least considered the possibility that the slaves were lying, but he did not. For the master, the "caste" distinction between healthy and diseased, between white and Negro, was overwhelming.

The tale of Brer Rabbit and the Tar Baby is well known today, appearing widely in children's books and in a novel by Toni Morrison

(*Tar Baby*, 1981). The version told in Jones (1888 [1969], pp. 7–11) goes like this: too lazy to find his own water, Rabbit steals from Wolf's spring. When Wolf tells Rabbit that he has seen Rabbit's tracks near his spring, Rabbit says they must have been from another rabbit. Doubtful, Wolf builds a tar baby and places it in the middle of the path to the spring. The next morning, Rabbit decides to go get some water from Wolf's spring to cool his burning cooking pot. He sees the tar baby and is astonished. He examines the tar baby closely and waits for it to move. The tar baby does not wink an eye, say anything, or move at all. Rabbit asks the tar baby to please move so he can get some water, but the tar baby doesn't answer. Rabbit asks again. Rabbit finally says, "Enty you know me pot duh bun? Enty you know me hurry? Enty you yeddy me tell you fuh move? You see dis han? Ef you dont go long and lemme git some water, me guine slap you ober." The tar baby still does not respond, and Rabbit slaps him on the head. Rabbit tries to pull his hand back and yells at the tar baby to let him go or else he will box him with his other hand. Rabbit's other hand gets stuck also. Rabbit continues to make threats, and since the tar baby never responds, Rabbit gets his knees and then his face stuck, and cannot pull loose. Wolf appears, declares that he has proved Rabbit's theft, and ties Rabbit to a bush and whips him with a switch. Rabbit hollers and begs, and finally asks Wolf to kill him instead by burning him up or knocking his brains out. Wolf says that kind of death would be too short, and so he will throw Rabbit into the briar patch, so the briars can scratch his life out. Rabbit says, "Do Buh Wolf, bun me: broke me neck, but dont trow me in de brier patch. Lemme dead one time. Dont tarrify me no mo." So Wolf throws Rabbit into the briar patch. Rabbit runs away saying, "Good bye, Budder! Dis de place me mammy fotch me up—dis de place me mammy fotch me up."

The tale ends in standard fashion, with Rabbit anticipating Wolf's action of throwing him in the briar patch and Wolf not considering whether Rabbit might be strategically lying. But Rabbit is not infallible, as shown by his altercation with the tar baby. The tar baby is strange and intriguing, to Rabbit as well as to us listeners. The tar baby is something between solid and liquid, between object and living being. If Wolf had simply poured out a puddle of tar for Rabbit to step in, or if Wolf had simply cornered Rabbit, most of the flavor of the tale would be lost. Rabbit's mistake is essential to the tale.

What exactly is Rabbit's mistake? Levine (1977, p. 115) says that the tale "underline[s] the dangers of acting rashly and striking out blindly." Smith (1997, p. 128) says that the tale "emphasizes that Br'er Rabbit can be duped by illusion but that he ultimately saves himself by remembering his 'home,' or cultural roots." Rabbit is indeed duped, but the tar baby is not an illusion: it is not intended to fool Rabbit's visual perception, as smoke, mirrors, or holograms might. Rabbit sees it just fine, is in fact astonished by its strangeness, and even examines it closely before addressing it. Rabbit is not at all rash; he takes time to examine the tar baby and waits for the tar baby to move before asking him. Rabbit does not strike out blindly; he first asks the tar baby to move, a quite normal social request, and even asks a second time. Rabbit sees the tar baby as a strange creature, but does not prejudge it and becomes angry only when the tar baby violates common courtesy.

Rabbit's mistake is that he thinks that the tar baby is a strategic actor. If masters do not see that pigeons and slaves are strategic actors, Rabbit's mistake is exactly the opposite: Rabbit thinks that everything is a strategic actor. Rabbit does not move the tar baby aside nor does he simply walk around it, the obvious courses of action if he thought that the tar baby were an object. Rabbit gets mad when the tar baby does not acknowledge his request to move aside. Rabbit does not attack the tar baby unconditionally but rather issues threats to the tar baby that he thinks a strategic actor would respond to. Rabbit even ascribes mental states and reasoning ability to the tar baby, saying that the tar baby should realize that Rabbit's pot is burning and that therefore Rabbit is in a hurry.

If these folktales teach the importance of strategic thinking, of recognizing that others are strategic actors, the tar baby tale cautions that one can overdo it; one can mistake objects for actors as well as mistake actors for objects. Hamilton (1985, p. 19) finds over three hundred versions of the tar baby tale, from Africa to India to the Bahamas to Brazil. One tar baby is strategic: in some parts of Georgia, the tar baby is a living monster who insults people and then traps them when they strike out at him in response.

The strategic folktale tradition continues with Richard Wright's parable of his fight with Harrison, and more recently with "The Saturday Morning Car Wash Club," a short story by James Ellis Thomas (2000) of Dothan, Alabama. Lorenzo agrees to help his friend Chester wash his car, a rusting brown 1978 AMC Pacer named Apollonia. However, Chester means to wash Apollonia not in his front yard but at the public car wash, so as to impress the girls who hang out there on Saturday mornings. The two sixteen-year-olds lurch into the car wash and step out, only to meet a smirking Leon and his three friends, who are three years older. Leon and his friends say there are no spots at the car wash for Chester's "piece of doo-doo." As the crowd gathers, expecting a fight, Lorenzo volunteers that Chester will race Leon for a spot. Back in the car, Chester, sobbing because his auntie gave him the car, cannot believe that Lorenzo challenged Leon on his behalf and just wants to go home.

Lorenzo asks Chester to lean his ear over so he can tell him what to do. Lined up toward the open road, the two cars gun their engines. There is a countdown and Leon speeds off on count "two." Chester doesn't move, and after Leon roars away, backs slowly into the now vacant car wash bay. "The hare had hauled tail and now the tortoise was taking up shack in the rabbit's hole. It was a Saturday morning cartoon." The crowd laughs and Le Ly, a girl Chester is sweet on, compliments Chester on his mirrored sunglasses.

As before, putatively weaker people triumph and gain real benefits through strategic thinking, by anticipating the actions of seemingly more powerful people obsessed with their own status. Lorenzo, who knows he is using "the oldest trick in the book," is better at strategic thinking than Chester, who is more concerned with appearances: how his hair looks and impressing the ladies. Chester is not good at understanding how others think; for example, he thinks his car is impressive and does not realize that everyone else considers it a rustmobile. Lorenzo credits his insight to childhood Saturday morning cartoons and, going a bit further back, folktales with tortoises and hares.

Compared to its well-known cultural and spiritual legacy, the strategic acumen of the 1960s civil rights movement in the United States is insufficiently appreciated (see also Hubbard 1968 and McAdam 1983). In January 1963, Wyatt Tee Walker presented to Southern Christian Leadership Council (SCLC) activists a detailed tactical plan to desegregate Birmingham, Alabama. He called it "Project C," for Confrontation (Branch 1988, p. 690; Williams 1987, p. 182). Walker's plan was strongly influenced by the movement's frustrations in its 1962 Albany, Georgia, campaign, its first attempt to mobilize an entire community to protest and fill the jails. In Albany, according to Charles Sherrod (1985), "Sometime we don't know who controls this, who controls the other. So we stomp around and stomp and see whose feet we get. . . . We didn't know what we were doing." Walker (1985) says that "we learned that valid and crucial lesson that you must pinpoint your target so that you do not dilute the strength of your attack." Also, the Albany chief of police, Laurie Pritchett (1985), had researched King's tactics beforehand and directed the Albany police to use "no violence, no dogs, no show of force." Sherrod and Walker dispute Pritchett's claim that he was nonviolent; as Sherrod (1985) observes, "How could a man be nonviolent who observed people being beaten with billy clubs." Walker (1985) instead uses the term "non-brutal," and explains, "the foil for our nonviolent campaigns in the South had been the uh, anticipated response of segregationist law enforcement officers such as Jim Clark, in Selma, and Bull Connor in Birmingham, Alabama. Laurie Pritchett was of a different stripe. . . . I think the apt description is slick. He did have enough intelligence uh, to read Dr. King's book, and he culled from that a way to avoid uh, confrontation."

Thus Project C focused on a single city block between 16th Street and 17th Street (Vann 1985, quoted in Williams 1987, p. 191) and on Birmingham's public safety commissioner, "Bull" Connor. Walker (1985) reminisces that Connor "served our purposes well.... [W]ithout the prototypical figure of a white racist law enforcement officer...the Birmingham Movement would not have accelerated and built up the momentum as fast. I often wonder why Bull Connor didn't have somebody smart enough around him to say, 'Let the niggers go on to City Hall and pray.'...He never had enough intelligence, or anybody around enough intelligence to let us do what we wanted to do. Instead he was fixed on stopping us, and that became the flash point of the dogs and the hoses and of the national and international attention in the 1964 Civil Rights Bills." Laurie Pritchett traveled to Birmingham to advise Connor, but they "never did agree on anything" (Pritchett 1985). Connor was a lame duck, having been recently voted out of office, and Walker knew that the window of opportunity was closing: Vann (1985) remembers Walker explaining that "they tried to talk us out of starting the demonstrations, and give the new government a chance. But we realized that this was our last chance, to demonstrate against Bull Connor.... [S]ooner or later he would do something that would help our cause." The movement's strategy of recruiting child demonstrators was due to James Bevel (1985, quoted in Williams 1987, p. 189), who explains that "a boy from high school, he get the same effect in terms of being in jail in terms of putting the pressure on the city as his father and yet he is not, there is no economic threat on the family because the father is still on the job." Even the police decision to use water hoses on the children was not so much a performative "show of force" but driven by necessity; with thousands of children already imprisoned for demonstrating (Bailey 1985) there was no space to imprison any more and thus the police had to "break up the demonstrations before they got started" (Walker 1985). Vann (1985) explains that "the ball game was all over, once the hoses and dogs were brought forward."

On May 11, 1963, a bomb exploded at the Gaston Motel, where Rev. Martin Luther King was staying. Another bomb exploded at the parsonage of Rev. A. D. King, Martin Luther King's brother. Pritchett had recommended that Connor guard the Gaston Motel with police, but Connor had refused. Wyatt Tee Walker's wife was also staying there with their children, and after the explosion, a state trooper at the scene "hit her with a carbine, split her head open, sent her to the hospital." After Walker arrived and was told which trooper hit his wife, Walker started toward him but "this white reporter from Mississippi,

Bob Gordon, tackled me and threw me to the floor and held me until I, you know, it occurred to me that...this guy would take this automatic rifle and shoot me as quickly as he had brained my wife....[I]t would have done irreparable harm to the non-violent movement. Because here was Dr. King's top lieutenant, chief of staff, attacking a...a police officer....I was committed to non-violence as a way of life, but for me there was no prohibition against me protecting my home and family. And with an attack, a physical attack on my wife, I guess in my mind, this guy was fair game and it didn't even occur to me, uh, you know, that he had an automatic weapon, you know. It's just one of those human responses....I'm very grateful that this white UPI reporter from Mississippi, stopped me" (Walker 1985). Walker, a central movement strategist, is not infallible. But pinned to the ground, he can think it through and is grateful.

This chapter sets the pattern for the rest of the book. Here we take African American slave folktales, understand them as analyses of strategic thinking, and consider the real-world applications of their insights. In the rest of the book, we do the same thing, first with *Flossie and the Fox*, and then with Austen's novels.

CHAPTER FOUR

Flossie and the Fox

THE FLOSSIE AND THE FOX story was told to Patricia C. McKissack (1986) by her grandfather. Flossie Finley, a little girl, is asked by her mother to deliver a basket of eggs to Miz Viola's place. Her mother warns her to watch out for the fox, who loves eggs. Flossie says that she doesn't know what a fox looks like; she doesn't remember ever seeing one. "Oh well, a fox be just a fox. That aine so scary." Flossie skips along and encounters a strange creature, who announces that he is a fox. Flossie looks him over carefully and says, "I just purely don't believe it.... I don't believe you a fox." Fox says that of course he is a fox: "A little girl like you should be simply terrified of me. Whatever do they teach children these days?" But Flossie replies, "I aine never seen a fox before. So, why should I be scared of you and I don't even-now know you a real fox for a fact?" Flossie goes on her way.

Fox, quite disconcerted, runs after Flossie and invites her to feel his thick fur. Flossie replies that he must be a rabbit. Fox then explains that he has a long pointed nose. Flossie replies that he must be a rat. After a while, they meet a cat, and Fox asks the cat to please explain to Flossie that he is indeed a fox. The cat says that he is a fox because he has sharp claws and yellow eyes, but Flossie concludes that he must therefore also be a cat. Desperately, Fox says that he has a bushy tail. Flossie replies that he must be a squirrel. Fox begs Flossie to believe him, but it is too late because one of Mr. McCutchin's hounds arrives to apprehend Fox. As he dashes away, Fox shouts that the hound knows who he is: "Like I told you, I am a fox!" Flossie replies, "I know," and walks unharassed to Miz Viola's.

There are several messages one can take from this story. One might say that Fox's terrifying power is based not on physical attributes but on socialization, what they teach children these days. One might say that the powerful construct a world with specific roles, and the weak can beat the powerful by refusing to participate in these roles; power requires acknowledgment, and disappears without it. One might say that Flossie succeeds by manipulating the situation, by steadfastly and cleverly refusing the fox and scared girl scenario in favor of the unknown creature and skeptical girl scenario; the real fight is not over the eggs but over how the situation is defined. One might say that whenever someone approaches you claiming to be powerful, you should place the burden of proof on him. One might say that childish innocence can triumph over adult pretension. One might say that Flossie simply knows how to buy time.

Anyhow, this story says something profound about the nature of power and resistance. The strategic insight here is that if Fox knows that Flossie can tell that he is a fox, then Flossie is at a disadvantage. Flossie gains not by being ignorant (after all, she reveals at the end that she knows he is a fox) but by making Fox think she is ignorant.

We can model this as a game in which Fox chooses whether to attack or not, and Flossie chooses whether to defend herself or not. If Fox does not attack, then nothing happens and the status quo is maintained. If Fox does attack and Flossie does not defend, then Fox gets the eggs without a fight; Flossie loses the eggs but at least there is no physical altercation. If Fox attacks and Flossie defends, then both Flossie and Fox risk injury. We represent Fox's and Flossie's preferences over these outcomes by numerical payoffs. Hence the game looks like this, where Flossie's payoffs and actions are in normal type and Fox's are in boldface.

TABLE 7

	Fox attacks	**Fox does not**
Flossie defends	−12, **−12**	0, **0**
Flossie does not	−8, **8**	0, **0**

Here the status quo in which Fox does not attack yields payoffs of 0 to both. If Fox attacks and Flossie does nothing, then Flossie loses the eggs (a payoff of −8) and Fox gets the eggs (a payoff of 8). If Fox attacks and Flossie defends, however, then both risk injury and both get a payoff of −12. The best thing for Fox is to steal the eggs without encountering any defense. The best thing for Flossie is to be left alone; losing the eggs is bad but being bitten or scratched is worse.

In this game, note that for Flossie, not defending is always at least as good as defending, regardless of what Fox does (−8 is greater than −12, and 0 is at least as great as 0). Thus we would expect Flossie not to defend the eggs. Given that Flossie does not defend, Fox gets a payoff of 8 if he attacks and a payoff of 0 if he does not. Thus we would expect Fox to attack.

But we have not yet captured the story, in which Fox is not sure whether Flossie knows that he is indeed a fox. Why should Fox care about whether Flossie knows he is a fox? The reason is that if Flossie thought that he were a squirrel, for example, then Flossie might act differently.

What would a game between Flossie and a squirrel look like? We write the game below.

TABLE 8

	Squirrel attacks	Squirrel does not
Flossie defends	0, −12	0, 0
Flossie does not	−8, 8	0, 0

The only difference from the earlier game (table 7) is that now Flossie does not incur any cost at all for defending if the squirrel attacks, since squirrels are small and beaten easily. Flossie's payoff if she defends and the squirrel attacks is now 0 (before it was −12). In this game, for Flossie, defending is always at least as good as not defending, regardless of what the squirrel does. Thus we would expect Flossie to defend the eggs. Given that Flossie defends, the squirrel gets a payoff of −12 if he attacks and 0 if he doesn't. Thus we expect the squirrel not to attack.

Of course, this game does not capture the story either. The situation in the story is a "blend" of these two games, in which Flossie's and Fox's knowledge about each other, and knowledge of each other's knowledge of each other, is crucial. This blend is called a "game with incomplete information." Since Flossie does not necessarily know if she is dealing with a fox or a squirrel, we say that she is dealing with a "creature." There are three relevant possibilities, or "states of the world," which Flossie and the creature must take into account: the creature is a fox and Flossie can tell, the creature is a fox and Flossie cannot tell, or the creature is in fact a squirrel. Flossie cannot distinguish between these last two states: if Flossie cannot tell, the creature could be a fox or a squirrel. The creature cannot distinguish between the first two states: the creature does not know if Flossie can tell or not. Of course, the creature does know whether he is a fox or a squirrel; he can distinguish between the third state and the other two.

To fully define the blended game, we must specify the probability of each state of the world. Assume that whether the creature is a fox or a squirrel is equally likely: the probability that the creature is a fox is 1/2 and the probability that the creature is a squirrel is 1/2. Assume that conditional on the creature being a fox, whether Flossie can tell or not is equally likely. So the probability of the first state of the world, that the creature is a fox and Flossie can tell, is 1/4. The probability of the second state of the world, that the creature is a fox and Flossie cannot tell, is 1/4. The probability of the third state of the world, that the creature is a squirrel, is 1/2.

In the blended game, Flossie chooses whether to defend or not in each of the three possible states of the world. But since Flossie cannot distinguish between the last two states of the world, she must take the same action in these two states. So Flossie has four possible strategies: (defend, defend, defend), (defend, not, not), (not, defend, defend), and (not, not, not). Here (defend, not, not), for example, means that Flossie defends in the first state but does not defend in the second and third (in other words, Flossie defends if she knows that the creature is a fox and does nothing otherwise). Note that (not, not, defend), for example, is not a possible strategy because Flossie must take the same action in the second and third states.

Similarly, the creature chooses whether to attack or do nothing in each of the three possible states of the world. The creature's possible strategies are (attack, attack, attack), (attack, attack, not), (not, not, attack), and (not, not, not). Since the creature cannot distinguish between the first two states, he cannot play for example (attack, not, attack).

Flossie has four strategies and the creature has four strategies. We noticed already that if Flossie knows that she is facing a fox, she will not defend, since defending is never better than not defending. So we can immediately eliminate Flossie's strategies (defend, defend, defend) and (defend, not, not), the strategies in which she defends when she knows that she is facing a fox. Thus Flossie has two remaining strategies: (not, defend, defend), in other words defend when she can't tell if it is a fox or squirrel, and (not, not, not), in other words never defend.

Thus the blended game is represented by the following table, which has two rows (Flossie's strategies) and four columns (the creature's strategies). Again, Flossie's payoffs and actions are in normal type and the creature's are in boldface.

TABLE 9

	(attack, attack, attack)	(attack, attack, not)	(not, not, attack)	(not, not, not)
(not, defend, defend)	$-5, -7$	$-5, -1$	$0, -6$	$0, 0$
(not, not, not)	$-8, 8$	$-4, 4$	$-4, 4$	$0, 0$

Note that if Flossie never defends and the creature always attacks, then Flossie always loses the eggs and gets -8, and the creature always gets the eggs and gets 8. If the creature never attacks, then both get 0 regardless of what Flossie does. The payoffs in this table are calculated using the probabilities of the three states of the world and the payoffs in the two original games (tables 7 and 8). For example, say Flossie plays (not, defend, defend) and the creature plays (attack, attack, not).

In the first state of the world, Flossie does not defend and the creature (a fox) attacks, and Flossie gets payoff −8 and the creature gets payoff 8. In the second state of the world, Flossie defends and the creature (a fox) attacks, and Flossie gets payoff −12 and the creature gets payoff −12 also. In the third state of the world, Flossie defends and the creature (a squirrel) does not attack, and both Flossie and the creature get payoff 0. Flossie's overall expected payoff is her payoff from each state multiplied by the probability of each state, summed up over all states. In other words, since she gets −8 in the first state (which occurs with probability 1/4), −12 in the second state (probability 1/4), and 0 in the third state (probability 1/2), her overall expected payoff is $(-8)(1/4)+(-12)(1/4)+(0)(1/2) = -5$. The creature's payoff is similarly $(8)(1/4) + (-12)(1/4) + (0)(1/2) = -1$. Hence when Flossie plays (not, defend, defend) and the creature plays (attack, attack, not), the entry in the table is −5, −1.

What will Flossie and the creature do in this game? The standard way to proceed (Nash's equilibrium [1950]) is by a process of elimination. Say, for example, that we predict that Flossie plays (not, not, not) and the creature plays (attack, attack, attack); in other words, Flossie never defends and the creature always attacks. This does not make much sense as a prediction, because if the creature always attacks, Flossie's payoff is −8 and she can do better by playing something different: she can get −5 by playing (not, defend, defend) instead. Given the prediction, Flossie does not want to play in a way consistent with the prediction. So the prediction of Flossie playing (not, not, not) and the creature playing (attack, attack, attack) does not make sense and is dropped. To take another example, say we predict that Flossie plays (not, defend, defend) and the creature plays (not, not, attack). Then the creature gets a payoff of −6 but could get a higher payoff of 0 by playing (not, not, not), never attacking. So this prediction also is dropped.

In a similar manner, one goes through the eight possible predictions and drops those in which at least one person could do better by not following the prediction. The one prediction which remains is that Flossie plays (not, defend, defend) and the creature plays (not, not, not); in other words, Flossie defends if she thinks the creature might be a squirrel, and the creature never attacks.

So in this blended game, we predict that the creature never attacks, even when Flossie knows he is a fox. In other words, when considering whether to attack, Fox must think about what Flossie might do. If Flossie cannot tell the difference between a fox and a squirrel, then Flossie will defend, thinking that she is quite possibly defending against a mere squirrel. Since Fox does not know whether Flossie can tell or not, he must consider the possibility of Flossie defending. This possibility is enough to deter Fox from attacking. This is true even when Flossie can tell that Fox is indeed a fox. This is how Flossie nullifies Fox's power by denying it recognition.

As is often noted, one problem with making threats (nuclear escalation, for example) is that they can be very costly to carry out and hence not credible. Thus the person making such a threat might want to give the impression that he is crazy enough to carry it out. Richard Nixon consciously employed this "madman theory" of deterrence (Schelling 1960 [1980]) when bombing Vietnam (Kimball 1998). Similarly, Flossie deliberately makes Fox think that she just might do something which is costly for herself. But the "Flossie theory" of deterrence is more sophisticated. Flossie introduces uncertainty in her opponent's mind, not about her sanity, but about whether she can recognize whether her opponent is powerful or weak. Flossie's uncertainty is more plausible and creative; anyone can pretend to be crazy.

In this chapter, we have told the same story twice: the first time as a folktale, the second time as a mathematical model (what we end up with in table 9). The advantages of the folktale are obvious: it is short, compelling, and easy for even children to understand. One might say that these advantages are exactly what the mathematical model does not have. The advantages of the model are less obvious. One advantage, very often exploited, is that a model makes it easy to ask "what if?" questions: for example, would Flossie's technique still work if the fox is starving and wants the eggs desperately? In the model, we can answer this question simply through computation: we would change the payoff numbers in table 7 and see what difference it makes to table 9. The main advantage of using mathematics, however, is simply that it is different; it has a different analytical momentum that can push one's analysis in unexpected directions (for example, the Beatrice-Benedick example and the Richard-Harrison example are both interesting taken separately, but we see their close similarity once we notate them technically, in tables 4 and 5). Mathematical models and narratives provoke and illustrate each other in a reciprocal fashion. Indeed, most successful game-theoretic arguments have both a mathematical representation and a compelling narrative that illuminates the argument in a simplified and stylized manner, like a folktale. Austen did not engage in game theory's mathematical development, but in chapter 14, I briefly argue that Austen might be understood as advocating this direction.

CHAPTER FIVE

Jane Austen's Six Novels

IN THIS CHAPTER, I survey Austen's six novels as chronicles of how a young woman learns strategic thinking skills, starting from as early as childhood. Strategic thinking not only helps you get married; learning strategic thinking is part of becoming a grown woman. When Fanny successfully manipulates her sister Betsey to give up Susan's knife, she is "fearful of appearing to elevate herself as a great lady" (MP, p. 459). Fanny is right to suspect that her manipulation of Betsey carries with it a change of status, but it is a necessary change, from childhood to womanhood, from girl to lady. Young women learn strategic thinking partly from reading novels and from watching their peers and older sisters, but mainly from having to make decisions themselves in demanding social situations.

All six novels discuss how a person learns strategic thinking; I take them in increasing order of the depth of their concern. I first discuss *Pride and Prejudice*, the "liveliest" novel but the one in which people's strategic abilities develop the least: Elizabeth Bennet and Mr. Darcy eventually recognize their mistakes but do not really acquire new strategic skills, having been well equipped from the start. *Sense and Sensibility* goes further, exploring through the sisters Elinor and Marianne Dashwood how strategic thinking requires both thoughtful decision-making (Elinor's strength) and fanciful speculation (Marianne's obsession). In *Persuasion*, the more mature (at twenty-seven) Anne Elliot also starts with strategic skills, but must learn to trust her own ability and outgrow her mother figure, Lady Russell. The next two novels explicitly describe how a young woman learns strategic thinking. In *Northanger Abbey*, seventeen-year-old Catherine Morland starts with no training but gradually learns strategic thinking by making decisions in a sequence of increasingly important situations. In *Mansfield Park*, Fanny Price, constrained and mistreated by her adopted family since the age of ten, must learn to make her own choices in the face of complete opposition. Finally, in *Emma*, Austen explores the dangers of learning too well, of being overconfident in your strategic ability.

The discussion here introduces many topics analyzed more systematically in the following chapters, such as the distinction between strategicness and selfishness, strategic partnership as the best foundation for marriage, making good choices even when overpowered by emotion, the necessity of understanding others' minds as different from your own, and the risk that status consciousness can make you strategically stupid. Here my main purpose is to summarize the novels, but sometimes analysis appears.

PRIDE AND PREJUDICE

For our purposes, *Pride and Prejudice* is Austen's most straightforward novel: Elizabeth Bennet and Mr. Darcy overcome their mutual disregard and realize their love for each other, but their strategic thinking skills do not substantially develop. Elizabeth is set up with strategic skills right from the start; according to her father, Mr. Bennet, she "has something more of quickness than her sisters" (PP, p. 5). Mrs. Bennet is eager to marry off her five daughters, and has designs on Mr. Bingley, the new tenant of Netherfield Park nearby. When Jane, the eldest, is invited by Mr. Bingley's sister Caroline to Netherfield, Mrs. Bennet has Jane go on horseback so that the anticipated rain would compel her to stay there all night and maximize face time with Mr. Bingley. What Elizabeth calls a "good scheme" works all too well and Jane falls ill, staying at Netherfield for several days (PP, p. 34).

Elizabeth visits her ailing sister and sees Mr. Darcy, Mr. Bingley's close friend. Earlier at a ball, Mr. Bingley had suggested that Mr. Darcy dance with Elizabeth, but Mr. Darcy made an unkind remark about her appearance, which Elizabeth overheard. Thus when Mr. Darcy asks her to dance at Netherfield, Elizabeth thinks he wants to humiliate her and declines, saying, "I always delight in overthrowing those kind of schemes" (PP, p. 56). Since Elizabeth "attracted him more than he liked," Mr. Darcy, "[s]teady to his purpose" and also strategically skilled, "wisely resolved to be particularly careful that no sign of admiration should *now* escape him" (PP, p. 66).

Elizabeth likes Mr. Darcy even less when the handsome Wickham tells her that Mr. Darcy's father had intended to bequeath to Wickham a living (a position as a local clergyman, which comes with a steady income), but Mr. Darcy ignored his father's wishes. Caroline Bingley warns Elizabeth that Wickham is not to be trusted, but Elizabeth sees Mr. Darcy's malice behind her intervention.

Since the Bennets have no son, Mr. Bennet's entailed property upon his death defaults to a male cousin, Mr. Collins, who shows up with the idea of marrying a Bennet daughter as a partial remedy. Mrs. Bennet hints that Jane might be engaged soon, and thus Mr. Collins proposes to Elizabeth, the second daughter. Mr. Collins doesn't take Elizabeth's no for an answer and appeals to Elizabeth's parents, but Mr. Bennet agrees with Elizabeth that Mr. Collins is a fool. The third Bennet daughter, Mary, thinks that "if encouraged to read and improve himself by such an example as her's, he might become a very agreeable companion," but Mr. Collins proposes successfully to Charlotte Lucas, Elizabeth's friend (PP, pp. 139–40).

When Mr. Bingley abruptly goes away to London, Jane concludes that he must not have had any real affection for her. Elizabeth suspects a conspiracy, however, and soon learns, indirectly through conversation with Mr. Darcy's cousin Colonel Fitzwilliam, that Mr. Darcy did indeed convince Mr. Bingley to stop pursuing Jane. Thus when Mr. Darcy unexpectedly proposes, Elizabeth responds angrily, blaming him for her sister Jane's unhappiness and for mistreating Wickham. The next day, Mr. Darcy writes a letter to Elizabeth explaining that he had not understood that Jane had truly cared for Mr. Bingley and that Wickham was deeply untrustworthy, having attempted to elope with his sister Georgiana Darcy when she was fifteen years old.

Elizabeth's uncle, Mr. Gardiner, and his wife, Mrs. Gardiner, offer to take her on a sightseeing trip northward, and Elizabeth agrees even though she is wary of going anywhere near Pemberley, Mr. Darcy's estate. Mrs. Gardiner has memories of Pemberley and wants to visit; Elizabeth is relieved at least that Mr. Darcy is scheduled to be away during their visit. Mr. Darcy returns home one day early and although their meeting is awkward, he treats Elizabeth and the Gardiners with the greatest kindness. Walking the grounds of Pemberley, Mrs. Gardiner, claiming fatigue, falls back with her husband, allowing Elizabeth and Mr. Darcy to walk alone together.

Jane writes Elizabeth with the alarming news that Wickham has run off with their younger sister Lydia, with no intention of marriage. After talking with the runaway couple in London, Mr. Gardiner writes to Mr. Bennet that they will marry as long as Mr. Bennet gives them Lydia's equal share of the five thousand pounds which the five Bennet daughters will receive upon the death of their parents, as well as an additional hundred pounds a year. Mr. Bennet, Elizabeth, and Jane are surprised that these amounts are so small and conclude that Mr. Gardiner himself must have paid off Wickham with a substantial sum, perhaps ten thousand pounds.

Elizabeth pries out of Mrs. Gardiner the information that Mr. Darcy first located Wickham and Lydia in London, and had settled the matter with his own wealth. Mr. Darcy had told Mr. Gardiner that he felt responsible because he did not warn anyone of Wickham's true character, but Mrs. Gardiner suspects that his true goal was Elizabeth's affection. Mr. Bingley arrives and happily proposes to Jane. Elizabeth, however, receives a surprise visit from the dour Lady Catherine de Bourgh,

Mr. Darcy's aunt. Lady Catherine, who had planned for Mr. Darcy to marry her own daughter, demands that Elizabeth promise to refuse any proposal from him. Elizabeth replies that she cannot make such a promise, infuriating Lady Catherine.

Mr. Darcy proposes to Elizabeth again, this time successfully, saying that he had gained hope when he heard from Lady Catherine that Elizabeth would not promise to refuse him. Mr. Darcy explains to Elizabeth that once he had told Mr. Bingley that he was convinced of Jane's affection for him, that was enough to make Mr. Bingley propose. Mr. Darcy is thankful for Lady Catherine, whose "unjustifiable endeavours to separate us, were the means of removing all my doubts" (PP, p. 423), and Elizabeth and Mr. Darcy are both thankful for the Gardiners, "who, by bringing her into Derbyshire, had been the means of uniting them" (PP, p. 431).

Throughout, Elizabeth's strategic ability remains constant; the reason that it is plausible that she, a very young woman "not one and twenty" (PP, p. 187), can react with such poise to Lady Catherine's aggressive surprise visit, in a manner decisive for her eventual marriage, is that she displays strategic quickness from the very beginning. Her strategic thinking draws her to reasonable but not always helpful conclusions; someone less strategic like Jane may have taken Caroline's warning about Wickham more literally, or may have taken Mr. Darcy's invitation to dance as an expression of interest, not contempt. Elizabeth's relationship to Mrs. Gardiner, a strategically skilled superior who looks out for Elizabeth's interests (in contrast to Elizabeth's own mother, Mrs. Bennet, a "woman of mean understanding, little information, and uncertain temper" [PP, p. 5]), does not change even after Elizabeth attains the status of married woman. Elizabeth can always use Mrs. Gardiner's help and is grateful for it.

Mr. Darcy had been certain that his first proposal would be accepted, and thereby learns the hard way that a proposal is a strategic situation: the proposer must consider whether the proposee will actually accept or not. Mr. Darcy acknowledges that Elizabeth's initial refusal was a "lesson" that "properly humbled" him (PP, p. 410). Otherwise, his strategic sense is pretty good from the very start also, even on matters between men and women. When Elizabeth first visits Netherfield to see Jane, Caroline asks Elizabeth to walk about the room together, and Mr. Darcy tells them that he can surmise two motivations: "You either chuse this method of passing the evening because you are in each other's confidence and have secret affairs to discuss, or because you are conscious that your figures appear to the greatest advantage in walking" (PP, p. 62). When Elizabeth and Mr. Darcy converse, she is "determined to leave the trouble of finding a subject to him" and subsequently Mr. Darcy "took the hint" (PP, p. 200). When Elizabeth refuses his first proposal, Mr. Darcy caustically retorts, "I thank you for explaining it so fully. My faults, according to this calculation, are heavy indeed!" (PP, p. 214). But Mr. Darcy should indeed thank Elizabeth for making her charges specific and thus enabling him to respond in full detail in his letter. Together Elizabeth and Mr. Darcy work together to solve a problem, and their existing strategic skills are sufficient for this.

Lydia Bennet comes out of nowhere to jump to the head of the marriage line, but she also does not change fundamentally in terms of her strategic skills, which were evident early on in her tracking of the regimental officers camped nearby and her "delicious scheme" of following them to Brighton (PP, p. 243). To Elizabeth and Jane, Lydia's marriage is an indecent reward for foolishness and licentiousness, but we might suspect that Lydia, perfectly aware of her own family's meager wealth, knows that her best shot at marrying with any money at all is to create a crisis situation in which wealthier members of the extended family, such as Mr. Gardiner, must bail her out to preserve the family reputation. Such a plan works only with a mercenary bridegroom not already committed to marrying her in the first place, and Wickham fills the bill. Lydia's display of utter recklessness also helps. Mrs. Bennet understands all of this perfectly, not blaming but rather incentivizing Lydia, telling her brother, Mr. Gardiner, that "as for wedding clothes, do not let them wait for that, but tell Lydia she shall have as much money as she chuses, to buy them, after they are married" (PP, p. 318). When Mr. Bennet is about to return home, giving up the rescue of Lydia to Mr. Gardiner, Mrs. Bennet complains that he is not playing his role as patriarchal enforcer: "Who is to fight Wickham, and make him marry her, if he comes away?" (PP, p. 329). After Wickham agrees to marry Lydia, Jane tries to shame her mother into gratitude by noting that Mr. Gardiner must have paid Wickham off, but Mrs. Bennet replies, "[I]t is all very right; who should do it but her own uncle?" (PP, p. 338).

One person who does develop her strategic skills is the remaining Bennet daughter Kitty, who doesn't understand her father's joke that she "has no discretion in her coughs" (PP, p. 6) and asks her mother, "What do you keep winking at me for? What am I to do?" when Jane and Mr. Bingley might be left alone together (PP, p. 382). Kitty soon learns that she would "rather stay at home" when asked if she wants to take a walk with Elizabeth and Mr. Darcy (PP, p. 416). Once Jane and Elizabeth are married, and Kitty spends time with them, "[i]n society so superior to what she had generally known, her improvement was great" (PP, pp. 427–28).

SENSE AND SENSIBILITY

Like Elizabeth Bennet, the Dashwood sisters Elinor and Marianne come
with fully functioning strategic skills in *Sense and Sensibility*. Elinor
"possessed a strength of understanding, and coolness of judgment" and
"Marianne's abilities were, in many respects, quite equal to Elinor's. She
was sensible and clever" (SS, p. 7). Elinor's skills do not develop much
in the course of the novel, but Marianne's require recalibration. Like
Brer Rabbit, who ascribes motivations to an inert object, Marianne is
much too eager to speculate about other people's motives, and eventually
learns how wrong she can be. Elinor is good at making decisions,
while Marianne overspecializes in thinking about other people's motives.
The two sisters thus exemplify two skills both necessary for strategic
thinking.
   Elinor, Marianne, and their youger sister Margaret grow up at
Norland, but when their father, Mr. Henry Dashwood, dies, the Norland
estate passes to Mr. John Dashwood, his son from a previous marriage.
Fortunately, their mother, Mrs. Dashwood, receives the offer of her
cousin Sir John Middleton to stay at Barton Cottage, nearby Barton Park,
where Sir John resides with his wife, Lady Middleton. At Barton, the
sisters' main task is to figure out what is going on with their uncertain
suitors. Marianne asks her mother why Edward Ferrars is pursuing
Elinor so slowly: "Twice did I leave them purposely together in the course
of the last morning, and each time did he most unaccountably follow
me out of the room" (SS, p. 47). Willoughby's pursuit of Marianne
is more rapid, and when Marianne excuses herself from a family
visit to Lady Middleton "under some trifling pretext of employment,"
Mrs. Dashwood, "who concluded that a promise had been made by
Mr. Willoughby the night before of calling on her while they were
absent, was perfectly satisfied with her remaining at home" (SS, p. 87).
When the family returns, finding Marianne in tears and Willoughby
not sure when he can call again, Elinor doubts her mother's belief
that Marianne and Willoughby are engaged. Mrs. Dashwood, who has
"eagerness of mind" like Marianne (SS, p. 7), thinks up a rationale for
Willoughby's sudden departure: Mrs. Smith, from whom he expects to
inherit, suspects his engagement with Marianne and disapproves, and
has therefore called him away. Mrs. Dashwood's belief in Willoughby is
based largely on his manner and appearance: "Has not his behaviour to
Marianne and to all of us, for at least the last fortnight, declared that
he loved and considered her as his future wife. . . . Has not my consent
been daily asked by his looks, his manner, his attentive and affectionate
respect?" (SS, p. 92). When Elinor suggests that her mother simply ask

Marianne if they are engaged, Mrs. Dashwood does not want to force Marianne's confidence and exclaims, "Supposing it possible that they are not engaged, what distress would not such an inquiry inflict!" (SS, p. 97). Elinor might be wary of her mother's and Marianne's fancifulness, but when Edward Ferrars also abruptly ends his visit with the Dashwoods for no apparent reason, Elinor also cannot stop thinking about Edward's true motivations, and must employ her own fancy: "the past and the future, on a subject so interesting, must be before her, must force her attention, and engross her memory, her reflection, and her fancy" (SS, p. 121).

Lucy Steele, a distant cousin of Lady Middleton, confides in Elinor that she has been secretly engaged with Edward for four years. Elinor easily understands Lucy's strategic objective in telling her so: "it required no other consideration of probabilities to make it natural that Lucy should be jealous; and that she was so, her very confidence was a proof. What other reason for the disclosure of the affair could there be, but that Elinor might be informed by it of Lucy's superior claims on Edward, and be taught to avoid him in future?" (SS, p. 162). Elinor and Marianne visit London with Mrs. Jennings, Lady Middleton's mother, and when Mrs. Jennings becomes occupied with the birth of her grandchild, Elinor and Marianne's half-brother John Dashwood proposes that they stay with his family. But his wife, Fanny Dashwood, is the sister of Edward Ferrars, and the Ferrars family plans for Edward to marry the rich Miss Morton. To prevent Elinor from getting any closer to Edward, Fanny replies that she had already invited Lucy Steele and her sister Anne Steele, of course not knowing that Lucy is a far greater risk.

Anne Steele, naively thinking that "they are all so fond of Lucy," reveals Lucy and Edward's secret engagement (SS, p. 293). Horrified, the Ferrars family ejects the Steele sisters and disowns Edward; Mrs. Ferrars, Edward's mother, installs his younger brother Robert as recipient of the Norfolk estate, which would have been Edward's. Now that the secret is out, Elinor explains Lucy's tactic to Marianne and her own strategic response: "It was told me,—it was in a manner forced on me by the very person herself, whose prior engagement ruined all my prospects; and told me, as I thought, with triumph.—This person's suspicions, therefore, I have had to oppose, by endeavouring to appear indifferent where I have been most deeply interested" (SS, pp. 298–99). Elinor's self-command makes this a teachable moment for Marianne, who had been reveling in unrestrained anguish over her mistreatment by Willoughby: "Oh! Elinor . . . you have made me hate myself for ever. . . . Because your merit cries out upon myself, I have been trying to do it away" (SS, p. 299).

Colonel Brandon, friend of the Dashwoods and admirer of Marianne, goes to Elinor and asks her to tell Edward that he would like to offer him a living on his estate, to help enable Edward, cut off from family support, to marry Lucy. Edward is grateful to Colonel Brandon but thinks Elinor truly responsible. Meanwhile, Marianne takes long thoughtful walks and catches cold from getting her shoes and stockings wet.

Marianne's health rapidly declines, and in a fever she cries out wildly for her mother. Elinor quickly decides to send Colonel Brandon to fetch Mrs. Dashwood, but Willoughby unexpectedly shows up first, having heard that Marianne is close to death. Willoughby seeks forgiveness, telling Elinor that he had felt true affection for Marianne, but that his benefactress Mrs. Smith had dismissed him for not marrying Colonel Brandon's very young niece Eliza Williams, with whom he had fathered a bastard child. With no money, he could not marry Marianne and thus had to marry the wealthy Miss Grey.

Mrs. Dashwood arrives and Marianne eventually recovers, vowing greater self-command: "[M]y feelings shall be governed and my temper improved. They shall no longer worry others, nor torture myself" (SS, p. 393). But she still can't help falling "back in her chair in hysterics" when she and her family learn that the former Lucy Steele now goes by the name Mrs. Ferrars, thus crushing any hope that Edward might still marry Elinor (SS, p. 400). But Edward shows up, and it turns out that Lucy managed to marry the newly endowed Robert Ferrars. Honorably released from his engagement, Edward proposes to Elinor, and Marianne, after much encouragement from Elinor and her mother, ends up happily with Colonel Brandon.

Elinor uses her strategic skills throughout. She convinces Marianne not to accept the gift of a horse from Willoughby early on in his supposed courtship, not by arguing for its impropriety ("She knew her sister's temper. Opposition on so tender a subject would only attach her the more to her own opinion") but by saying that the horse would inconvenience their dear mother (SS, p. 69). Later she makes sure to intercept news of Willoughby's wedding so that she can break it to Marianne gently, "desirous that Marianne should not receive the first notice of it from the public papers" (SS, p. 246). After she learns of Edward's engagement to Lucy, she tries to "weaken her mother's dependence on the attachment of Edward and herself, that the shock might be less when the whole truth were revealed" (SS, p. 179), and similarly tries to keep Mrs. Jennings from perceiving Marianne's fervent hopes that Willoughby will call on them in London. Most importantly, Elinor persuades Edward, who is rightfully upset, to reconcile with his mother, which results in her consent to their marriage and his reinclusion into the family, as well as ten thousand pounds.

Of course, Lucy Steele is equally strategic: when Robert Ferrars privately visits her to convince her to drop her engagement with Edward,

Lucy parlays his visit into several recurring visits, demonstrating how "an earnest, an unceasing attention to self-interest, however its progress may be apparently obstructed, will do in securing every advantage of fortune" (SS, p. 426). After Edward and Elinor are engaged, Edward still wants to think that Lucy had true affection for him, because she still wanted to marry him even after his family disowned him. Elinor replies that Lucy perhaps anticipated that a crisis situation would motivate the help of others: "[S]he might suppose that something would occur in your favour; that your own family might in time relent" (SS, p. 416). In other words, perhaps Lucy's tactic was similar to Lydia Bennet's in *Pride and Prejudice*. After all, Lucy earlier wrote Elinor asking "to recommend him to any body that has a living to bestow" (SS, p. 315), and Colonel Brandon ended up offering one. For that matter, Lucy had previously anticipated that Robert would be the correct target: when she and Elinor first met, she told Elinor that she and Edward did not dare mention their engagement to Mrs. Ferrars because "in her first fit of anger upon hearing it, [she] would very likely secure every thing to Robert" (SS, p. 169). Lucy and Elinor are different not in their strategic skills but in their objectives. Strategic skills do not have to be used for mercenary purposes, such as marrying money and quashing competitors. One can be a strategic Elinor without being a gold digger Lucy.

Understanding other people's motives requires an active imagination, but Marianne's is hypertrophic. For example, Colonel Brandon haltingly tells Elinor that he once knew a woman like Marianne who suffered unfortunate circumstances. For Elinor, "it required but a slight effort of fancy to connect his emotion with the tender recollection of past regard. Elinor attempted no more. But Marianne, in her place, would not have done so little. The whole story would have been speedily formed under her active imagination" (SS, p. 67). Mrs. Jennings idly remarks that hunters are out in the countryside but will return as winter advances. Marianne, wondering why the hunter Willoughby has not yet called on them in London, "saw every night in the brightness of the fire, and every morning in the appearance of the atmosphere, the certain symptoms of approaching frost" (SS, p. 191). Of course, Marianne's worst misconception is thinking that she was actually engaged to Willoughby: "I felt myself . . . to be as solemnly engaged to him, as if the strictest legal covenant had bound us to each other." When Elinor replies that "unfortunately he did not feel the same," Marianne submits, "He *did* feel the same, Elinor—for weeks and weeks he felt it. I know he did" (SS, p. 214).

Overconfidence in your ability to know others' true motives can lead to solipsism. When Mrs. Jennings hands a letter to Marianne, Marianne's "imagination placed before her a letter from Willoughby, full of tenderness and contrition" (SS, p. 229). But crushingly, the letter turns out to be from Mrs. Dashwood instead, and Marianne concludes that Mrs. Jennings, who had only tried to comfort, had intentions of the greatest cruelty. Marianne "expected from other people the same opinions and feelings as her own, and she judged of their motives by the immediate effect of their actions on herself" (SS, p. 229). Given the "too great importance placed by her on the delicacies of a strong sensibility," Marianne believes that others have the same sensibility as herself and should therefore see the effects of their actions like she does (SS, pp. 228–29). Anyone whose actions give her pain should be able to see that and is thus doing so on purpose. Marianne comes dangerously close to resembling the patently ridiculous Mrs. Palmer, Mrs. Jennings's daughter, who believes that she would have been married to Colonel Brandon, who is a close friend of Sir John Middleton, "[b]ut mama did not think the match good enough for me, otherwise Sir John would have mentioned it to the Colonel, and we should have been married immediately" (SS, p. 135). Mrs. Palmer imagines this even though she had met Colonel Brandon only twice.

Yet without speculation, fancy, and imagination, strategic thinking is impossible. For example, when Elinor convinced Marianne not to accept Willoughby's present of a horse by appealing to their mother's convenience, she did so based on her expectation that Marianne would respond contrarily if she more directly argued for its impropriety. But this expectation, however well grounded, could have been incorrect, and cannot be verified short of going back in time and seeing what would have happened had Elinor tried that strategy instead. Anyone in a social situation must conjecture about the motivations of other people and how they will respond to one's actions. If one waits until these conjectures are somehow made certain, one would not act at all. As Mrs. Dashwood notes, "Are no probabilities to be accepted, merely because they are not certainties?" (SS, p. 91).

Marianne's strategic skills can be quite advanced. When Elinor and Marianne are invited to the home of Mrs. Ferrars, a pair of screens painted by Elinor are passed around for admiration. But when the screens are remarked to be in the style of Miss Morton, a rare moment of appreciation for Elinor threatens to morph into an all-out Miss Morton praise session, and thus Marianne bursts "into tears.—Every body's attention was called, and almost every body was concerned" (SS, p. 269). Later, after being dumped by Willoughby, Marianne wanted to return home immediately to see her mother, but had to wait for months; only when she is sick and cries out wildly in a fever, terrifying Elinor, is Mrs. Dashwood actually sent for. Are these emotional outbursts purposeful? In both instances, Marianne objectively succeeds:

further praise of Miss Morton is prevented, and she finally gets to see her beloved mother. It also must be said that Marianne's incessantly displayed cultivation of delicacy makes both paroxysms more believable. The big payoff, however, is Willoughby's rush to Marianne, to ask for her forgiveness and declare that his affection had been genuine. When Elinor receives him and asks why he came, Willoughby explains, "What I felt on hearing that your sister was dying—and dying too, believing me the greatest villain upon earth, scorning, hating me in her latest moments. . . . My resolution was soon made, and at eight o'clock this morning I was in my carriage" (SS, pp. 374–75).

Would Marianne actually allow herself to become ill, even risking death, just to motivate Willoughby's journey of repentance or at least hasten her mother's arrival? She caught cold by taking long walks where "the grass was the longest and wettest" and her illness was "assisted by the still greater imprudence of sitting in her wet shoes and stockings" (SS, p. 346). After she recovers, Marianne admits to Elinor, "My illness, I well knew, had been entirely brought on by myself by such negligence of my own health, as I had felt even at the time to be wrong. Had I died,—it would have been self-destruction" (SS, p. 391). Getting sick is another way to create a crisis situation. But perhaps we are making the same mistake as Marianne did with Mrs. Jennings, seeing strategic premeditation where none exists. Perhaps Marianne is not engaging in brinkmanship but is truly going mad.

Is Marianne at risk for schizophrenia? People with schizotypy (proneness to schizophrenia) are more likely to ascribe mental states to "entities generally perceived to lack mental capacities" such as trees (Gray, Jenkins, Heberlein, and Wegner 2011, p. 478), and indeed when leaving the Norland estate, Marianne bids it goodbye: "Oh! happy house, could you know what I suffer in now viewing you from this spot, from whence perhaps I may view you no more!" (SS, p. 32). In this sense, people with schizotypy are the opposite of people on the autistic spectrum, who are less likely to ascribe mental states to people. Another aspect of schizotypy is "difficulties of attention and concentration" (Nettle and Clegg 2006, p. 612), and indeed "[a] surprising amount of the narrative tension of *Sense and Sensibility* comes from the bent bow of the absentation of Marianne's attention from wherever she is" (Sedgwick 1991, pp. 827–28). Two other aspects of schizotypy are "perceptual and cognitive aberrations and magical thinking" and "impulsive nonconformity . . . violent and reckless behaviors"; the first is correlated with artistic creativity and both are correlated with mating success, measured in number of sexual partners (Nettle and Clegg 2006, p. 612).

At the same time, and not necessarily in contradiction, people who commit suicide (in the United States at least) are predominantly male and "usually plan the act, take precautions to avoid interruption, and chiefly use rapidly effective, generally irreversible means. Their purpose is to die . . . and the great majority succeed on the first attempt." However, much more common (perhaps ten times more) are "parasuicides." People who commit parasuicide are predominantly female, "make provisions for rescue (others present or notified), and employ slowly effective or ineffective means. Their purpose is to survive and (usually) to send a message to another person" (Murphy 1998, p. 166).

Thirteen-year-old Margaret, the youngest Dashwood sister, starts out strategically naive. Early on, when Mrs. Jennings tries to get Margaret to reveal the name of Elinor's suitor, "Margaret answered by looking at her sister, and saying, 'I must not tell, may I, Elinor?' This of course made every body laugh" (SS, p. 71). But near the end when Edward Ferrars visits, after the Dashwood family has understood him to have married Lucy Steele, "Margaret, understanding some part, but not the whole, of the case, thought it incumbent on her to be dignified, and therefore took a seat as far from him as she could, and maintained a strict silence" (SS, p. 407). Margaret's strategic abilities are developing, and soon she "reached an age highly suitable for dancing, and not very ineligible for being supposed to have a lover" (SS, p. 431).

PERSUASION

At twenty-seven years of age, Anne Elliot is the oldest of Austen's heroines, and like Elizabeth Bennet and the Dashwood sisters, comes ready-made with strategic skills. But *Persuasion* is a coming-of-age story in that Anne must outgrow her superior. Anne comes to realize that her dear friend Lady Russell, the intimate of her late mother who has been guiding her since childhood, has not helped her make the best decisions. In the end, "[t]here was nothing less for Lady Russell to do, than to admit that she had been pretty completely wrong, and to take up a new set of opinions and of hopes" (P, p. 271). Lady Russell loves Anne and has done her best to counsel Anne in her most important decision, whom to marry, but Anne comes to "learn that she and her excellent friend could sometimes think differently" (P, p. 159). Despite Lady Russell's greater life experience, "[s]he was a woman rather of sound than of quick abilities" (P, p. 12). Anne is simply better at strategic thinking. As opposed to Elizabeth Bennet, who remains "always on the most intimate terms" with Mrs. Gardiner (PP, p. 431) in *Pride and Prejudice*, and as opposed to the Dashwood sisters, whose relationship with their mother doesn't change upon marriage in *Sense and Sensibility*, Anne has to outgrow Lady Russell, and their relationship must be reestablished on new terms, with her husband, Captain Wentworth, valuing Lady Russell "in spite of all her former transgressions" (P, p. 274).

When Anne was nineteen, she had received Captain Wentworth's proposal, but Lady Russell persuaded Anne against it. At the time, Captain Wentworth had little going for him but attitude: "He was brilliant, he was headstrong.—Lady Russell had little taste for wit; and of any thing approaching to imprudence a horror" (P, p. 29). Eight years later, having succeeded in the navy, Captain Wentworth returns to visit his sister Mrs. Croft, who with her husband, Admiral Croft, is renting Kellynch Hall, the estate of Sir Walter Elliot, Anne's father. Sir Walter had to rent out the estate because of his financial profligacy, and has moved to cheaper quarters in Bath with his eldest daughter Elizabeth and her widow friend Mrs. Clay, leaving Anne behind with Lady Russell and Anne's other sister Mary Musgrove. Mary is married to Charles Musgrove, whose grown sisters are Louisa and Henrietta Musgrove.

Passive among the Musgroves and Captain Wentworth, "Anne's object was, not to be in the way of any body" (P, p. 90). Mostly Anne tries to perceive Captain Wentworth's feelings: "to retentive feelings eight years may be little more than nothing. Now, how were his sentiments to be read?" (P, p. 65). Anne is mortified when she hears that Captain Wentworth finds her "[a]ltered beyond his knowledge"; while playing the piano, "she felt that he was looking at herself—observing her altered features, perhaps, trying to trace in them the ruins of the face which had once charmed him" (P, pp. 65, 77–78). To understand him, Anne puts herself in his place: "Had he wished ever to see her again, he need not have waited till this time; he would have done what she could not but believe that in his place she should have done long ago" (P, p. 63). Anne is obsessed enough that she begins to presume that she understands him better than anyone else: "There was a momentary expression in Captain Wentworth's face . . . a certain glance of his bright eye, and curl of his handsome mouth . . . but it was too transient an indulgence of self-amusement to be detected by any who understood him less than herself" (P, p. 73).

But stronger evidence than words or glances are Captain Wentworth's actions, which demonstrate his own strategic acumen. First he pries Mary's two-year-old son off Anne's back; as opposed to Anne and her neighbor Charles Hayter, Captain Wentworth knows that two-year-olds cannot be "ordered, intreated," or incentivized; sometimes you just have to grab them (P, p. 86). Next, during a long walk, the Musgroves, Charles Hayter, Anne, and Captain Wentworth run into Admiral and Mrs. Croft in their carriage. There is room in the carriage for one other person, and the Crofts offer to take one of the young women home. But "Captain Wentworth cleared the hedge in a moment to say something to his sister," resulting in Mrs. Croft insisting that Anne go with them (P, p. 97). Captain Wentworth knows that Anne is too modest to accept the offer made to the four young ladies as a group, but cannot decline such a direct offer. Anne understands all this, and "was very much affected by the view of his disposition towards her, which all these things made apparent.... She understood him. He could not forgive her,—but he could not be unfeeling.... [H]e could not see her suffer, without the desire of giving her relief" (P, p. 98).

Despite Anne's passivity, her investment of energies into detection as opposed to action, when an emergency arises she is best at making decisions. The group visits Captain Wentworth's friend Captain Harville at Lyme and are taking a seaside walk on the Cobb, a stone harbor wall. Louisa has so much enjoyed having Captain Wentworth help her jump down the steep steps of the Cobb that she wants to do it again, despite his reservations that the pavement is too hard. "[S]he smiled and said, 'I am determined I will:' he put out his hands; she was too precipitate by half a second, she fell on the pavement on the Lower Cobb, and was taken up lifeless!" (P, p. 118). Captain Wentworth, "as if all his own strength were gone," cries out, "Is there no one to help me?" (P, p. 118). Anne first responds and directs the group to use smelling salts, rub Louisa's temples and hands, and call for a surgeon. When Captain Wentworth sets off for the surgeon, Anne points out that Captain Benwick, a friend who has been staying with Captain Harville, should go instead since he is the only one in the party who knows Lyme well. The women, Mary and Henrietta, are insensible or hysterical, and the remaining men, Charles Musgrove and Captain Wentworth, "[b]oth seemed to look to her for directions. 'Anne, Anne,' cried Charles, 'what is to be done next? What, in heaven's name, is to be done next?'" (P, pp. 119–20).

Anne directs them to take Louisa to the inn, but Captain Harville and his wife, informed by Captain Benwick running by, appear and direct the party to their home, where the surgeon pronounces Louisa not hopeless. Anne returns to the Musgrove residence with Henrietta and Captain Wentworth, but is still in charge, persuading Louisa's parents to stay in lodgings at Lyme until Louisa recovers, and arranging for the Musgrove family's nursery maid to nurse Louisa. Louisa gradually improves and the rescue party disperses, with Anne going with Lady Russell to join her family in Bath, as originally planned, and Captain Wentworth off to see his brother.

In Bath, Anne finds that Mr. William Walter Elliot, the cousin who is the heir presumptive to Sir Walter's estate, has ingratiated his way into the family, overcoming his offenses several years earlier: instead of marrying Anne's sister Elizabeth, as the family had hoped, he had married another wealthy woman, without consulting Sir Walter, head of the house, and had even spoken contemptuously of the family. Mr. Elliot, now a widower, visits Sir Walter and Elizabeth regularly; Elizabeth's widow friend Mrs. Clay is also still hanging around. Anne soon finds herself his target, and Lady Russell recommends that Anne accept his addresses, hoping that Anne will be restored as Lady Elliot: "[O]ccupying your dear mother's place, succeeding to all her rights, and all her popularity, as well as to all her virtues, would be the highest possible gratification to me" (P, p. 173).

The Crofts arrive in Bath, ostensibly to relieve the Admiral's gout, and Anne receives the surprising news that Louisa Musgrove is marrying Captain Benwick. Now that Louisa is no longer eligible for Captain Wentworth, Anne "had some feelings which she was ashamed to investigate. They were too much like joy, senseless joy!" (P, p. 182).

When Captain Wentworth arrives in Bath, Anne finally takes subtle but crucial strategic actions. Sitting inside a shop on Milsom Street as it starts to rain, Anne is startled to see through the window Captain Wentworth walking down the street. Anne manages to manipulate herself into going toward the door and Captain Wentworth unexpectedly enters the shop and bumps into her. Later, Anne attends a concert because she knows that Captain Wentworth likes music; she works up the courage to address him if he arrives, by saying to herself that she owes him the attention, since her sister Elizabeth had pointedly turned away from him at Milsom Street. Captain Wentworth enters, and "making yet a little advance, she instantly spoke. He was preparing only to bow and pass on, but her gentle 'How do you do?' brought him out of the straight line to stand near her" (P, p. 197). After their conversation, they lose each other in the crowd, and the widower Mr. Elliot annoyingly finds a seat next to Anne as the concert begins. During intermission, Captain Wentworth does not present himself, but through "a little scheming of her own, Anne was enabled to place herself much nearer the end of the bench than she had been before, much more within reach of a passer-by. . . . [S]he found herself at the very end of the bench before the concert closed" (P, p. 206). Captain Wentworth brusquely wishes Anne good night and rushes off, forcing Anne to conclude, "Jealousy of Mr. Elliot! It was the only intelligible motive. . . . For a moment the gratification was exquisite. But alas! there were very different thoughts to succeed. . . . How, in all the peculiar disadvantages of their respective situations, would he ever learn her real sentiments?" (P, p. 207).

Captain Harville and members of the Musgrove family also surprisingly appear in Bath, reuniting the Lyme company, excepting Louisa and Captain Benwick. When Anne visits them at their hotel, she finds Mrs. Musgrove (the mother of Charles, Louisa, and Henrietta) talking with Mrs. Croft, and Captain Harville talking with Captain Wentworth.

Captain Harville asks to speak with Anne, and with Captain Wentworth within listening distance, Captain Harville laments the fickleness of Captain Benwick, who had been engaged to his sister Fanny, but after her death is so quickly marrying Louisa: "Poor Fanny! she would not have forgotten him so soon!" (P, p. 252). Anne agrees, and Captain Harville guides the conversation into a comparison of whether men or women are more constant in their attachment. Through gentle but successively greater challenges, Captain Harville goads Anne into speaking increasingly warmly of women's constancy, with Anne finally ending with the declaration: "I should deserve utter contempt if I dared to suppose that true attachment and constancy were known only by woman. No, I believe you capable of every thing great and good in your married lives.... All the privilege I claim for my own sex (it is not a very enviable one, you need not covet it) is that of loving longest, when existence or when hope is gone" (P, p. 256). Overhearing this cry, Captain Wentworth is moved to silently write a letter to Anne declaring the constancy of his own heart. He exits, but under the pretense of recovering his gloves, comes back in and gives the letter to Anne without attracting notice.

Anne is so overpowered by the letter that she does not understand what people around her are saying. Listening comprehension skills might momentarily desert her but her strategic ability does not: she immediately refuses the offer to be carried home by a chair so that she will not lose the possibility of catching up with Captain Wentworth as he walks through town, makes a backup plan by asking Mrs. Musgrove twice to promise to assure Captains Harville and Wentworth that they are both expected at her father's party that evening, and even makes a backup backup plan, assuring herself that if Captain Wentworth doesn't show, she can send word to him via Captain Harville. Charles Musgrove volunteers to walk Anne home, but when they meet Captain Wentworth in the street, Charles remembers an engagement of his and asks Captain Wentworth to serve in his stead, with all happiness resulting.

What had brought Captain Wentworth to Bath was Anne's revealed preference. He knew that she had refused the proposal of Charles Musgrove three years after his own: "It was possible that you might retain the feelings of the past, as I did; and one encouragement happened to be mine. I could never doubt that you would be loved and sought by others, but I knew to a certainty that you had refused one man at least, of better pretensions than myself: and I could not help often saying, Was this for me?" (P, p. 265).

By marrying Captain Wentworth, Anne declares her independence from Lady Russell's advice, now eight years old. The final strike against Lady Russell's judgment was that the widower Mr. Elliot turned out to be "a man without heart or conscience; a designing, wary, cold-blooded being, who thinks only of himself; whom for his own interest or ease, would be guilty of any cruelty, or any treachery, that could be perpetrated without risk of his general character" (P, p. 215). These are the words of Anne's old school friend Mrs. Smith, living impoverished in Bath due to the negligence of Mr. Elliot, who was a close friend of her deceased husband and the executor of his will (note that Mr. Elliot's cruelty is not an inherent characteristic but a choice he makes with an awareness of its possible costs). Anne had long suspected Mr. Elliot's character and realizes that she escaped Lady Russell just in time: "Anne could just acknowledge within herself such a possibility of having been induced to marry him, as made her shudder at the idea of the misery which must have followed. It was just possible that she might have been persuaded by Lady Russell!" (P, pp. 228–29). In the end, Lady Russell accepts her diminished role: "[S]he was a very good woman, and if her second object was to be sensible and well-judging, her first was to see Anne happy. She loved Anne better than she loved her own abilities" (P, pp. 271–72).

By outgrowing Lady Russell, Anne is not cast out alone in an atomistic world; to find and secure love, sometimes you need all the help you can get. Who helps Anne?

Anne had supposed that apart from her own immediate family and Lady Russell, only one other person had known about Captain Wentworth's first proposal, his brother Edward, because his sister, Mrs. Croft, had been out of the country at the time. But surely Anne should suspect that in the course of eight years, most likely his proposal had been talked about, quite possibly by Captain Wentworth himself; the only reason why Anne thinks that Edward has not talked about it is that he was a single man (and therefore conclusively untalkative) at the time. Before the Crofts rented Sir Walter's estate, Mrs. Croft knew that her brother Captain Wentworth knew Anne. When Anne first meets her, Anne feels she is on safe ground until Mrs. Croft suddenly asks, "It was you, and not your sister, I find, that my brother had the pleasure of being acquainted with, when he was in this country.... Perhaps you may not have heard that he is married?" (P, pp. 52–53). Anne recovers once Mrs. Croft explains that she is talking about her brother Edward, not Captain Wentworth. It is hard to imagine a better way for Mrs. Croft to gauge Anne's visceral interest in her unmarried brother and at the same time give Anne space to recover. Mrs. Croft recommends the married naval life to a group including Anne (and Captain Wentworth) and "always met her with a kindness which gave her the pleasure of fancying herself a favourite" (P, p. 136). Of course, Mrs. Croft knows what to do when her brother asks her to entreat Anne to accept a ride with her and Admiral Croft in their carriage. When Anne hears the Crofts talk together in the carriage about Captain Wentworth's marriage hopes, Mrs. Croft's tone when discussing the Musgrove sisters "made Anne suspect that her keener powers might not consider either of them as quite worthy of her brother" (P, p. 99). After Captain Benwick and Louisa are set to marry, Admiral Croft tells Anne that Captain Wentworth "must begin all over again with somebody else. I think we must get him to Bath. Sophy [Mrs. Croft] must write, and beg him to come to Bath.... Do not you think, Miss Elliot, we had better try to get him to Bath?" (P, p. 188). Even Edward inquires "very particularly" about Anne (P, p. 264).

Charles Musgrove, who had proposed to Anne before succeeding with her sister Mary, is continually throwing some single man in Anne's direction, courting her by proxy. Charles tells Lady Russell that Captain Benwick spoke of Anne using the words "Elegance, sweetness, beauty" and will surely visit them soon, although he never does (P, p. 142). When Charles, Captain Harville, and the various other Musgroves show up in Bath, the motivation for their trip, and whether Captain Wentworth is with them, is mysterious, and even Mary, usually obsessed only with herself, is in on it: "Anne could not draw upon Charles's brain for a regular history of their coming, or an explanation of some smiling hints of particular business, which had been ostentatiously dropped by Mary, as well as of some apparent confusion as to whom their party consisted of" (P, p. 234). Charles proposes that they all go see a play ("I have engaged Captain Wentworth. Anne will not be sorry to join us, I am sure") instead of attending Sir Walter's party to which Mr. Elliot is invited (P, p. 242). Charles's proposal is not really serious, but it does allow Anne to state in the presence of Captain Wentworth that if it were up to her she would prefer the play over the party.

When Anne visits the Musgroves at their hotel, where Captain Wentworth will soon write his letter, little does she know that she is walking into an ambush. Mary and Henrietta have just cleared out, leaving Mrs. Musgrove, Mrs. Croft, and Captains Harville and Wentworth. Anne is told that "the strictest injunctions had been left with Mrs. Musgrove, to keep her there till they returned" and therefore she must sit there, captive along with Captain Wentworth, the other target (P, p. 249). Mrs. Musgrove and Mrs. Croft, "in that inconvenient tone of voice which was perfectly audible while it pretended to be a whisper," agree that long engagements are best avoided and it is always better to marry at once (P, p. 250). Anne "felt its application to herself, felt it in a nervous thrill all over her, and at the same moment that her eyes instinctively glanced towards the distant table, Captain Wentworth's... head was raised, pausing, listening, and he turned round the next instant to give a look—one quick, conscious look at her" (P, p. 251). Captain Harville calls Anne to stand with him nearby Captain Wentworth, and induces her proclamation of woman's eternal constancy. Predictably inspired, Captain Wentworth writes his letter to Anne and leaves with Captain Harville. After Anne reads the letter, Charles Musgrove, with an excuse preset ("an engagement at a gunsmith's"), volunteers to escort Anne home, only to exit immediately upon meeting Captain Wentworth, leaving the two together (P, p. 260). Anne had "felt almost certain of meeting him" in the street (P, p. 259), but objectively speaking, a chance meeting was unlikely since Captain Wentworth had left several minutes before her. Captain Wentworth had left in the care of Captain Harville, and Anne had left in the care of Charles Musgrove, and Harville and Musgrove are the two who organized the trip to Bath together in the first place: "The scheme had received its first impulse by Captain Harville's wanting to come to Bath on business. . . . Charles had proposed coming with him" (P, p. 235).

Are Anne and Captain Wentworth guided from start to finish? Are Mrs. Croft, Captain Harville, and Charles Musgrove to Anne what Mrs. Gardiner (in *Pride and Prejudice*) is to Elizabeth Bennet? Austen encourages us to trace everyone's motivations and actions in detail, to exercise our own strategic thinking.

NORTHANGER ABBEY

Catherine Morland in *Northanger Abbey* does not come equipped with strategic skills: "[s]he never could learn or understand any thing before she was taught; and sometimes not even then, for she was often inattentive, and occasionally stupid" (NA, p. 6). Strategic skills are not inborn but must be taught. Catherine receives instruction from real-world decision-making situations, as well as from peers and novels.

Given that her older siblings are all brothers, Catherine starts out with understandably little instruction in strategic thinking. Catherine at the age of ten loves boys' plays and was "noisy and wild, hated confinement and cleanliness" and even at fourteen loves "cricket, base ball, riding on horseback, and running about the country" (NA, pp. 6–7). But to become a grown woman, she must learn how to think strategically. Her real training begins once she visits Bath, at the age of seventeen, accompanying her neighbors Mr. and Mrs. Allen, who have no children of their own to take with them.

At Bath, Catherine is befriended by Isabella Thorpe, four years older and her presumed strategic superior: Isabella "could discover a flirtation between any gentleman and lady who only smiled on each

other.... These powers received due admiration from Catherine, to
whom they were entirely new" (NA, p. 26). Isabella tells Catherine
that she prefers light eyes and sallow complexion in a man, and that
"[y]ou must not betray me, if you should ever meet with one of
your acquaintance answering that description" (NA, p. 36). Catherine,
not understanding, replies, "Betray you!—What do you mean?" (NA,
p. 36). When Catherine asks Eleanor Tilney whether she thinks the girl
her brother Henry Tilney danced with is pretty, Catherine reveals her
own feelings for Henry "without the smallest consciousness of having
explained them" (NA, p. 71). Catherine is "not experienced enough in
the finesse of love" and is not yet skilled at reading others' feelings: when
her own brother James Morland and Isabella's brother John Thorpe
unexpectedly arrive in Bath, James greets Isabella "with a mixture of
joy and embarrassment which might have informed Catherine, had she
been more expert in the development of other people's feelings, and less
simply engrossed by her own, that her brother thought her friend quite
as pretty as she could do herself" (NA, pp. 29, 39).

Catherine's strategic naivety is due to a lack of training in her
own family. In contrast, the Thorpe family, which produced Isabella,
encourages fancy, surmises, and indirection. After James proposes to
Isabella, Catherine thinks it unkind that Isabella's younger sisters are
not told directly, "but Anne and Maria soon set her heart at ease by the
sagacity of their 'I know what;' and the evening was spent in a sort of
war of wit, a display of family ingenuity, on one side in the mystery of
an affected secret, on the other of undefined discovery, all equally acute"
(NA, p. 123).

Catherine begins her training in earnest when she is placed in a
sequence of increasingly consequential situations. The first is when she
must choose whether to join her brother James, Isabella, and John
Thorpe on their country drive, even though she had originally planned
to look for Eleanor Tilney and her brother Henry Tilney at the pump
room. Catherine excitedly chooses to go on the drive, but afterward
when Mrs. Allen recounts that Henry and Eleanor were indeed at the
pump room and had even conversed with her for half an hour, Catherine
learns that one should always look ahead: "Could she have foreseen such
a circumstance, nothing should have persuaded her to go out with the
others" (NA, p. 66). The next evening, at the cotillion ball, Catherine
must figure out how to be available to dance with Henry Tilney but at the
same time avoid John Thorpe: "she fidgetted about if John Thorpe came
towards her, hid herself as much as possible from his view, and when he
spoke to her pretended not to hear him" (NA, p. 72). At the ball, Henry
and Eleanor and Catherine make a plan to go on a walk the next day at
noon as long as it does not rain. At eleven it starts to rain lightly, and at

half past twelve, it starts to clear. But James, Isabella, and John Thorpe unexpectedly show up and ask Catherine to join them on another drive, this time to Blaize Castle. Catherine must predict whether the Tilneys will keep the engagement, and must think about their preferences: "whether there had not been too much rain for Miss Tilney to venture, must yet be a question" (NA, p. 82). For the first time Catherine thinks of the Tilneys as making a choice; earlier, Catherine regarded the Tilneys' appearance at the pump room as a matter of luck, a circumstance. John Thorpe reports that he saw the Tilneys riding off in a carriage, and Catherine concludes that "I suppose they thought it would be too dirty for a walk"; since Catherine has a romantic interest in castles, the matter is settled and the party rides off (NA, p. 84). On the way, Catherine spots the Tilneys walking down the street and entreats John Thorpe to stop. Thorpe instead speeds his horses forward and it is apparent to Catherine that Thorpe had deceived her. Now Catherine must somehow make amends to the Tilneys; when she sees Henry the next evening at the theatre, her eager apology is more than successful, and they schedule their walk again. But again the riding party asks her to drop her engagement and ride with them instead, and this time Catherine resolutely refuses. John Thorpe then goes to Eleanor Tilney, telling her that Catherine had sent him to say that she had been previously engaged for the ride, and thus cannot go with the Tilneys. After John Thorpe proudly returns to report his crude manipulation, Catherine runs after the Tilneys with even greater resolve, declaring, "If I could not be persuaded into doing what I thought wrong, I never will be tricked into it" (NA, p. 101).

Catherine, who originally was so easily persuadable, who could even be flattered into saying that she likes John Thorpe, has experienced the downside of going with the flow and has now been toughened up enough to make her own decisions given her own preferences. After James proposes to Isabella, and John Thorpe asks her, "Did you ever hear the old song, 'Going to one wedding brings on another?'... [W]e may try the truth of this same old song," Catherine is more than prepared to reply, "May we?—but I never sing" (NA, p. 125).

Does Catherine learn strategic thinking from her peers? Isabella continually imputes strategicness to Catherine, but only to advance her own interests. Isabella tells Catherine that when she met her brother James, they had discovered that "our opinions were so exactly the same, it was quite ridiculous!... [Y]ou are such a sly thing, I am sure you would have made some droll remark or other about it.... You would have told us that we seemed born for each other, or some nonsense of that kind" (NA, p. 68). Isabella informs Catherine about James's marriage proposal by saying that Catherine must have figured it out already; Catherine cannot "refuse to have been as full of arch penetration and affectionate sympathy as Isabella chose to consider her" and thereby Isabella, concerned about Catherine's parents' consent given her own family's lack of fortune, gains Catherine's reassurance (NA, p. 121). Mr. and Mrs. Morland happily consent, and Mr. Morland promises to give his own living, worth four hundred pounds a year, in a few years to James, as soon as he is old enough. When Isabella hints that this is insufficient ("every body has their failing, you know, and everybody has right to do what they like with their own money"), Catherine is somewhat hurt (NA, p. 138). Isabella neutralizes Catherine by appealing to her penetration: "Ah! my Catherine, you have found me out. There's the sting. The long, long, endless two years and half that are to pass before your brother can hold the living." This ploy works and "Catherine's uncomfortable feelings began to lessen. She endeavoured to believe that the delay of the marriage was the only source of Isabella's regret" (NA, p. 139).

Isabella thus exposes Catherine to strategic manipulation, but in the end Isabella is most instructive for showing how it can degenerate into the ludicrous. The engagement is broken off because Isabella openly flirts with and even plans to be engaged to Eleanor's other brother, Captain Tilney. But after Captain Tilney disappears, Isabella writes Catherine asking after James, saying that "I...am afraid he took something in my conduct amiss." Catherine's judgment is finally unequivocal: "Such a strain of shallow artifice could not impose even upon Catherine. Its inconsistencies, contradictions, and falsehood, struck her from the very first....Her professions of attachment were now as disgusting as her excuses were empty" (NA, pp. 223–24).

Catherine learns more from Henry Tilney, who is around twenty-five. Henry knows all too well the importance of knowing the minds of others. When they first meet, Henry jokingly but repeatedly asks what Catherine is thinking: he asks what she will write in her journal about their first meeting, and even asks directly, "What are you thinking of so earnestly?" Catherine thinks "that he indulged himself a little too much with the foibles of others" (NA, p. 21). During their second dance a few days later, Henry remarks that matrimony and dancing are strategically similar and indeed can be represented by the same game tree: "[I]n both, man has the advantage of choice, woman only the power of refusal" (NA, p. 74). Perhaps "Henry's clever chaste comparison of marriage to a country dance reflects the novel's sassy sexlessness" (Brownstein 1997, p. 38), but right now Henry is concerned with the strategic, not the sexual. Henry humorously notes that dancing and marriage are both contracts ("when once entered into, they belong exclusively to each other till the moment of its dissolution"), and when he asks whether Catherine will again interrupt their dance by talking with John Thorpe or anyone else, Catherine, properly prodded into a strategic mindset, does not reply in terms of etiquette or propriety but in terms of feasibility: "[T]here are hardly three young men in the room besides him, that I have any acquaintance with" (NA, pp. 74, 75). Henry is only satisfied, however, once Catherine replies in terms of preferences: "I do not *want* to talk to any body" (NA, p. 76).

Later, when Catherine ascribes Captain Tilney's wish to dance with Isabella to his good nature, Henry replies in theoretical terms, "How very little trouble it can give you to understand the motive of other people's actions.... With you, it is not, How is such a one likely to be influenced? What is the inducement most likely to act upon such a person's feelings, age, situation, and probable habits of life considered?— but, how should *I* be influenced, what would be *my* inducement in acting so and so?" (NA, pp. 134–35). As a general statement about the tendency to suppose others' preferences as similar to your own, and the need to instead understand how they can be quite different, Henry could not be more explicit.

Novels also play a role in Catherine's training: "[F]rom fifteen to seventeen she was in training for a heroine; she read all such works as heroines must read" (NA, p. 7). The strategically minded Isabella Thorpe and Henry Tilney, as well as Eleanor Tilney, are also accordingly novel readers, but the less fanciful Mrs. Morland, Catherine's mother, does not keep up with the newest ones and the strategic imbecile John Thorpe declares, "Novels are all so full of nonsense and stuff" (NA, p. 43).

General Tilney, Henry and Eleanor's father, invites Catherine to return with them to their home at Northanger Abbey. In this new setting, Catherine is quick to use the strategic skills she learned in Bath: when she receives the letter from her brother James saying that his engagement to Isabella is off, she does not mention it directly to Henry and Eleanor, since it concerns their brother Captain Tilney. Instead, "she might just give an idea—just distantly hint at it" (NA, p. 209). She asks Eleanor and Henry, "I have one favour to beg...that, if your brother should be coming here, you will give me notice of it, that I may go away." When Henry figures out what happened, Catherine cries, "How quick you are!...[Y]ou have guessed it, I declare!—And yet, when we talked about it in Bath, you little thought of its ending so" (NA, p. 210). Catherine thus makes clear to Henry that she was right when earlier she had asked Henry to call off his brother for the sake of her own, and that at least in this case she bests her tutor at predicting others' actions.

Catherine has improved at acting and predicting, but her weakness still is in figuring out others' motivations (in contrast to Anne Elliot, who is preoccupied over others' feelings but is slow to act). For example,

General Tilney says that he is tempted to order a new set of breakfast china, and hopes "that an opportunity might ere long occur of selecting one—though not for himself. Catherine was probably the only one of the party who did not understand him," and only somewhat later does she finally figure out that the General hopes that she and Henry will wed (NA, p. 179).

What speculative capacities Catherine has are very much overinfluenced by the romantic novels she has read. General Tilney seems anxious that Eleanor not show Catherine the room where her mother died, and according to Eleanor, Mrs. Tilney's portrait is hung in her own bedchamber and not in the drawing room because General Tilney was not satisfied with it. Catherine quickly suspects General Tilney of having imprisoned his poor wife in hidden regions of the abbey. Henry, having read the same novels, with "his quick eye fixed on hers," immediately understands her suspicions, and admonishes, "Remember that we are English, that we are Christians. Consult your own understanding, your own sense of the probable, your own observation of what is passing around you—Does our education prepare us for such atrocities? Do our laws connive at them? Could they be perpetrated without being known, in a country like this, where social and literary intercourse is on such a footing; where every man is surrounded by a neighbourhood of voluntary spies, and where roads and newspapers lay every thing open?" (NA, p. 203). Henry's argument is not that his father or the English or Christians have fine moral character, but simply that a person who would do such a thing would get caught (by "spies" Sutherland and Le Faye [2005, p. 156] suggest that Henry means neighbors and servants). Henry is telling Catherine that she still has to work on her strategic reasoning.

General Tilney leaves for London for a week, and Catherine, Eleanor, and Henry are happy to be free from the demands of his presence: "His departure gave Catherine the first experimental conviction that a loss may be sometimes a gain" (NA, p. 227). Learning from peers or novels is fine, but the best lessons are real-world "experimental" ones (according to Knox-Shaw [2004, p. 17], Austen's "empirical habit of mind" is due to her early exposure to the sciences). When General Tilney returns, however, he immediately orders Catherine sent home, to Eleanor's horror (Henry is not present, having left for a few days to his curate at Woodston). When Eleanor comes to tell Catherine of her expulsion, Catherine, chastened by her earlier error into restraining her fancy and "always judging and acting in future with the greatest good sense" (NA, p. 206), errs in the other direction, underestimating the General's rudeness, until Eleanor mortifyingly reveals that Catherine will be sent away unaccompanied by a servant and at a time not of her choosing. The excuse is that General Tilney had recalled a previous engagement, a ploy as crude as John Thorpe's earlier.

Catherine returns home safely, and although the Morland family acknowledges the General's ill-treatment of their daughter, "[w]hy he had done it, what...so suddenly turned all his partial regard for their daughter into actual ill-will...did not oppress them by any means so long" (NA, p. 242). Mrs. Morland calls the whole thing a learning experience: "Catherine is safe at home, and our comfort does not depend upon General Tilney....It is always good for young people to be put upon exerting themselves; and you know, my dear Catherine, you always were a sad little scatter-brained creature; but now you must have been forced to have your wits about you" (NA, p. 243). Speaking to Mrs. Allen, Mrs. Morland adds, "And it is a great comfort to find that she is not a poor helpless creature, but can shift very well for herself" (NA, p. 246). Mrs. Morland is referring only to Catherine's unaccompanied carriage ride home, but her remark more properly describes all Catherine has learned in Bath and Northanger Abbey since leaving home.

Still, Catherine wonders if she will ever see Henry again, and Mrs. Morland, who is generally not good at understanding Catherine's feelings, attributes her languor to her missing her grand Northanger lifestyle. But when Henry appears, saying that he wanted to make sure that Catherine made it home safely, Mrs. Morland notices Catherine's "glowing cheek and brightened eye" and is savvy enough to allow Catherine to show him the way to the Allens' house (NA, p. 251). Walking together, Henry proposes and Catherine accepts. Henry explains that General Tilney had invited Catherine to Northanger thinking that she was the heiress of the wealthy and childless Allens, because John Thorpe, who was hoping to marry Catherine himself, told him so at Bath. When the General was away at London, he met with John Thorpe again, who, "under the influence of exactly opposite feelings, irritated by Catherine's refusal, and yet more by the failure of a very recent endeavour to accomplish a reconciliation between [James] Morland and Isabella," told him this time that the Morlands were "a necessitous family...a forward, bragging, scheming race" (NA, pp. 255–56). Under the influence of her novels, Catherine had been only partially wrong: "in suspecting General Tilney of either murdering or shutting up his wife, she had scarcely sinned against his character, or magnified his cruelty" (NA, p. 256). Of course, Henry has not obtained his father's consent to the marriage, and Catherine has learned enough to appreciate Henry's strategicness in proposing to her first: she "could not but rejoice in the kind caution with which Henry had saved her from the necessity of a conscientious rejection, by engaging her faith before he mentioned the subject" (NA, p. 253). But after Eleanor marries well, General Tilney finally consents.

More than Austen's other heroines, Catherine knows what she wants and takes steps, right from the start, to get it. She goes after Henry without worrying much about whether he loves her. Catherine could be better at understanding the motives of others, but this does not stop her from making plans and pursuing him.

Catherine receives lots of help. Of course, most of the interaction between Catherine and Henry would not be possible without the presence and mediation of Eleanor Tilney as friend and sister. The Allens are most useful: they take Catherine to Bath in the first place, give her false consequence in the minds of John Thorpe and hence General Tilney, and at the end provide a destination for Henry and Catherine's walk together. But little of this is intentional, and in fact Mrs. Allen is sadly lacking in strategic skills. When she takes Catherine to her first dance at Bath, "Mrs. Allen did all that she could do in such a case by saying very placidly, every now and then, 'I wish you could dance, my dear,—I wish you could get a partner.' For some time her young friend felt obliged to her for these wishes; but they were repeated so often, and proved so totally ineffectual" (NA, p. 14). Mr. Allen, a "sensible, intelligent man," is better, ascertaining that Elizabeth's dance partner, Henry Tilney, is "a clergyman, and of a very respectable family in Gloucestershire" (NA, pp. 12, 22). General Tilney, like Lady Catherine in *Pride and Prejudice*, is unintentionally instrumental: "the General's unjust interference, so far from being really injurious to their felicity, was perhaps rather conducive to it, by improving their knowledge of each other, and adding strength to their attachment" (NA, p. 261). Behind General Tilney's actions are the oscillating exaggerations of John Thorpe, who was also angling to marry into money. As in the Brer Rabbit tale, others' mistaken schemes can be your best opportunities.

Catherine's younger sister Sarah is sixteen and curious about the strategic world. Upon Catherine's return, she wonders about General Tilney: "I can allow for his wishing Catherine away, when he recollected this engagement...but why not do it civilly?" But this is the Morland family, not the Thorpe family, and Mrs. Morland replies, "My dear, you give yourself a great deal of needless trouble...depend upon it, it is something not at all worth understanding" (NA, p. 243). When Henry Tilney asks Catherine to show him the way to the Allens' house, Sarah, trying to be helpful, volunteers, "You may see the house from this window, sir," only to receive a "silencing nod" from Mrs. Morland (NA, p. 252). Sarah has yet to read novels and take her own course at Bath.

MANSFIELD PARK

In *Mansfield Park*, Fanny Price's strategic development is distilled into just two crucial decisions. The first is whether to join her cousins and their friends in acting in a play at their home, despite her belief that the head of the family, her uncle Sir Thomas Bertram, would surely disapprove. The second is whether to accept Henry Crawford's proposal of marriage. The first decision is not important in itself, but Fanny agonizes over it and steadfastly refuses. Fanny exercises her power of choice and thus this first refusal is a rehearsal for her second refusal, which dumbfounds everyone, given Henry's wealth and agreeableness and Fanny's financial and, seemingly, intellectual dependence. In neither case is Fanny's steadfastness infinite; in the end Fanny agrees to read a part in the play, and Henry would have succeeded had he been even more persistent. But Fanny, raised from childhood to be subservient, unconfident, and voiceless in her own household, learns to make her own decisions with admirable, and in the end fruitful, resolve. Fanny is introduced to strategic thinking by the antics of the young people around her. She never becomes a strategist at even the basic level of Catherine Morland in *Northanger Abbey*, but she does not need to. Sometimes you don't have to be good at reading others' minds or thinking a few moves ahead; you just have to make good choices.

Fanny's development is traced from the age of ten. Fanny's aunts, Lady Bertram and Mrs. Norris, take pity on their relatively impoverished sister Mrs. Price, who has a ninth child on the way, and offer to raise Fanny, Mrs. Price's eldest daughter. Fanny arrives at Mansfield Park, home of Sir Thomas and Lady Bertram, their two sons, Tom and Edmund, ages 17 and 16, and their two daughters, Maria and Julia, ages 13 and 12.

Displaced and despondent, Fanny is received by a household concerned with specifying her feelings. Sir Thomas stresses how crucial it is "to preserve in the minds of my *daughters* the consciousness of what they are, without making them think too lowly of their cousin; and how, without depressing her spirits too far, to make her remember that she is not a *Miss Bertram*" (MP, p. 12). Mrs. Norris is occupied with what Fanny should feel as opposed to what she does feel. "Mrs. Norris had been talking to her the whole way from Northampton of her wonderful good fortune, and the extraordinary degree of gratitude and good behaviour which it ought to produce, and her consciousness of misery was therefore increased by the idea of its being a wicked thing for her not to be happy" (MP, p. 14). The exception in the household is Edmund, who sees Fanny crying on the stairs. Edmund finds that Fanny misses her family but especially her brother William; Edmund helps her write a letter to him, and "[f]rom this day Fanny grew more comfortable. She felt that she had a friend" (MP, p. 19). True kindness, and even basic sociality, requires understanding a person's wants. Edmund is "always true to her interests, and considerate of her feelings" (MP, p. 24).

Edmund's thoughtfulness extends to strategic reasoning. When Fanny is sixteen, her grey pony dies and a replacement is not thought necessary. Mrs. Norris warns that Sir Thomas would not approve the additional expense and thinks it "absolutely unnecessary, and even improper, that Fanny should have a regular lady's horse of her own in the style of her cousins." Edmund thus finds a "method of proceeding which would obviate the risk of his father's thinking he had done too much, and at the same time procure for Fanny the immediate means of exercise," by trading one of his three horses for a mare suitable for Fanny (MP, p. 42). Edmund is Fanny's tutor: "he recommended the books which charmed her leisure hours, he encouraged her taste, and corrected her judgment" (MP, p. 25).

Dr. and Mrs. Grant reside at the parsonage nearby, and Mrs. Grant's half-siblings Henry and Mary Crawford visit. Mary is interested in learning how to ride a horse, and Edmund offers to teach her on Fanny's mare. Mary learns so quickly that one suspects she already knows how: "Where people wish to attach, they should always be ignorant.... A woman especially, if she have the misfortune of knowing any thing, should conceal it as well as she can" (NA, p. 112). Fanny, watching Edmund and Mary together, "felt a pang.... [I]f she were forgotten the poor mare should be remembered," and indeed, for all excepting Edmund, Fanny does not rank far above an animal in consideration (MP, p. 79). When Fanny feels ill and retreats to the sofa, having worked for hours in the sun cutting and delivering roses for her aunts, Edmund realizes his neglect. "[S]he had been left four days together without any choice of companions or exercise, and without any excuse for avoiding whatever her unreasonable aunts might require," and Edmund "very seriously resolved...that it should never happen again" (MP, p. 87). Thus Fanny happens upon the technique that Marianne Dashwood successfully used in *Sense and Sensibility* to draw in Willoughby's repentance.

Maria, now twenty-one, is engaged to the wealthy but stupid Mr. Rushworth, and they await Sir Thomas's return from Antigua to get married. When Mr. Rushworth talks at length about improving his family's estate, Sotherton, Mrs. Norris suggests that they should all visit to suggest improvements. Since Lady Bertram is not going, Mrs. Norris insists that Fanny remain to keep her company, but when Edmund offers to remain at home instead so that Fanny can go, Mrs. Norris says that she had told Mrs. Rushworth, Mr. Rushworth's mother, that Fanny would not be coming. Edmund, anticipating this objection, had already secured Mrs. Rushworth's invitation for Fanny and thus Mrs. Norris is foiled. Mrs. Grant, who had earlier suggested to her brother Henry Crawford that he might marry Julia Bertram, knows that her sister Mary has parallel designs on Edmund, and thus offers to remain with Lady Bertram so that Edmund can go after all. Mrs. Norris's original scheme is thus twice amended with the schemes of others. Henry and Mary Crawford, Edmund, Maria, Julia, Mrs. Norris, and finally Fanny set off for Sotherton, with Mrs. Grant offering Julia the seat next to Henry, leaving Maria "in gloom and mortification" (MP, p. 94).

At Sotherton, Fanny, Edmund, and Mary Crawford walk the grounds and sit together on a bench near an iron gate to rest. Mary Crawford wants to start moving again, and although Fanny feels fully rested also, Edmund entreats her to rest some more, saying that he and Mary will be back in a few minutes. Fanny waits alone for twenty minutes until Mr. Rushworth, Maria, and Henry show up. The iron gate is locked but Maria still wishes to go through it, and thus Mr. Rushworth is compelled to go back to the house to get the key. In his absence, Maria and Henry go around the gate, again leaving Fanny behind. Now Julia arrives, having finally managed to shake off Mrs. Rushworth, and speeds off toward Maria and Henry. When Mr. Rushworth arrives with the key, Fanny has to say that Maria and Henry proceeded without him. Fanny offers, "Miss Bertram thought you would follow her," but this is not enough to make the upset Mr. Rushworth get up from the bench. Mr. Rushworth walks up to the gate and "Fanny thought she discerned in his standing there, an indication of relenting" (MP, p. 119). Thus she appeals to his status as the decider: "It is a pity you should not join them. They expected to have a better view of the house from that part of the park, and will be thinking how it may be improved; and nothing of that sort, you know, can be settled without you." This second attempt succeeds. Mr. Rushworth replies, "[I]f you really think I had better go; it would be foolish to bring the key for nothing" (MP, p. 120). Mr. Rushworth's making a decision based on costs already incurred is an example of the "sunk cost fallacy" (Friedman, Pommerenke, Lukose, Milam, and Huberman 2007; according to Elster [2007, p. 218], another example is the U.S. reluctance to disengage from Vietnam).

This is the first time Fanny purposefully takes an action to influence the action of another, based on her understanding of what motivates him. It is so mild as to be almost nothing and has no objective greater than simple consideration for Mr. Rushworth, but it is still a step in learning strategic reasoning: "Mr. Rushworth was worked on" (MP, p. 120).

Fanny, passive on the bench, twice observes how couples manage to drop their third; back at Mansfield Park, Fanny tries out what she has learned. As Fanny, Edmund, and Mary Crawford talk together at the window, the Miss Bertrams invite Mary to come accompany their singing. Fanny, now alone with Edmund, rhapsodizes about the night's beauty, and suggests, "I wish I could see Cassiopeia" (MP, p. 132). As Knox-Shaw (2002, p. 45) observes, "[V]isible at Mansfield Park only from the garden lawn, as Fanny at her window well knows, the hidden constellation provides just the pretext for drawing Edmund out of the gravitational field of Mary Crawford." It almost works. Edmund suggests going out to the lawn to see the stars together, but is diverted by the music, and Fanny "sighed alone at the window" (MP, p. 133).

Tom Bertram had gone abroad with Sir Thomas but returns home first, and suggests that Mansfield put on a play. Edmund and Fanny are certain that Sir Thomas would disallow it if he were home, but Tom and the others are undeterred, and after endless discussion choose Lovers' Vows. As to the casting, when Tom's friend Mr. Yates offers to play either Baron Wildenhaim or Frederick, and Henry Crawford also doesn't mind playing either role, Maria notes that as the tallest, Mr. Yates would be best as the Baron. Thus "she was certain of the proper Frederick" (MP, p. 156). Henry Crawford reciprocates by arguing that Maria, not Julia, would be best as Agatha, who shares several scenes with Frederick. Julia "saw a glance at Maria which confirmed the injury to herself; it was a scheme—a trick; she was slighted, Maria was preferred; the smile of triumph which Maria was trying to suppress shewed how well it was understood" (MP, p. 157). Henry suggests that Julia should play Amelia, but Tom insists that Mary Crawford would be better for Amelia. Fanny observes all this thrusting and counterthrusting and Julia storms out of the room, refusing to take part.

Two roles are yet uncast: Anhalt and Cottager's wife. Anhalt plays opposite Amelia, and Mary Crawford hopefully calls out, "What gentleman among you am I to have the pleasure of making love to?" (MP, p. 169). Mary goes as far as explicitly suggesting Edmund, since Anhalt is a clergyman and Edmund will himself soon take orders. Cottager's wife, which Mr. Yates calls "the most trivial, paltry, insignificant part," falls to Fanny (MP, p. 158). Tom Bertram does not ask as much as command: "Fanny . . . we want your services. . . . You must be Cottager's wife" (MP, pp. 170–71). The others press in on Fanny: "the requisition was now backed by Maria and Mr. Crawford, and Mr. Yates . . . which altogether was quite overpowering to Fanny" (MP, p. 172). Mrs. Norris horrifically declares, "I shall think her a very obstinate, ungrateful girl, if she does not do what her aunt and cousins wish her—very ungrateful indeed, considering who and what she is" (MP, p. 172).

Fanny goes to bed, "her nerves still agitated by the shock of such an attack from her cousin Tom, so public and so persevered in, and her spirits sinking under her aunt's unkind reflection and reproach. To be called into notice in such a manner...and then to have the charge of obstinacy and ingratitude follow it, enforced with such a hint at the dependence of her situation.... [W]hat should she do?" (MP, p. 176). In the morning she goes to the East room, which was once the family schoolroom, where Fanny and the Miss Bertrams had taken lessons as young children. Abandoned by the Miss Bertrams, its educative function remains for Fanny, as the decision she makes there about whether to join the play is an important step in her strategic development. She takes her first step away from pliability and ductility, but if anything she is too careful and self-reflective. She tests her own motivations and also considers strategically whether Edmund will back her up: "Was she *right* in refusing what was so warmly asked, so strongly wished for? What might be so essential to a scheme on which some of those to whom she owed the greatest complaisance, had set their hearts? Was it not ill-nature—selfishness—and a fear of exposing herself? And would Edmund's judgment, would his persuasion of Sir Thomas's disapprobation of the whole, be enough to justify her in a determined denial in spite of all the rest?" (MP, p. 179).

Edmund interrupts Fanny's deliberations, asking for help with a decision of his own. The company is on the verge of inviting a complete outsider to play Anhalt, and Edmund feels that he must play the part himself, in order to prevent a stranger from gaining rapid intimacy with their family and further exposing their folly. Edmund also asks Fanny to think how Mary Crawford would feel to play Amelia with a stranger. Edmund implores, "Give me your approbation, then, Fanny. I am not comfortable without it," but he doesn't allow Fanny to fully respond, as his decision is already made. When Fanny does not immediately approve, Edmund states, "I thought *you* would have entered more into Miss Crawford's feelings" (MP, p. 182). What point is there in Fanny's careful self-reflection when her reference and guide is so befuddled? "Could it be possible? Edmund so inconsistent. Was he not deceiving himself? Was he not wrong? Alas! it was all Miss Crawford's doing. She had seen her influence in every speech, and was miserable" (MP, pp. 183–84). All of Edmund's carefulness and consistency bends to the prospect of dramatic lovemaking (which Fanny is forced to witness, unbearably).

Fanny has from the age of ten relied on Edmund completely: he "formed her mind." But Fanny begins to wonder about his judgment when Mary Crawford arrives: "there began now to be some danger of dissimilarity, for he was in a line of admiration of Miss Crawford, which might lead him where Fanny could not follow" (MP, p. 76). When Fanny hints to Edmund that Henry Crawford seems to admire the engaged Maria more than Julia, Edmund thinks there is nothing to be concerned about, and Fanny takes this as wisdom; she "supposed she must have been mistaken, and meant to think differently in future." But Fanny's doubts linger, and "she knew not always what to think" (MP, p. 137). When the idea of a play is first brought up, Fanny hopes that the group will have great difficulty finding a play acceptable to all; Edmund is less optimistic, but indeed it takes days for the group to agree, and "Fanny seemed nearer being right than Edmund had supposed" (MP, p. 153). Edmund's choice to act in the play is final proof of his fallibility. Edmund is the pliable one, not Fanny. Like Anne Elliot in *Persuasion*, who realizes the limitations of her guardian Lady Russell, Fanny must learn to trust her own judgment. Like Catherine Morland in *Northanger Abbey*, who gains resolve after learning that John Thorpe tried to trick her into riding with him, Fanny more strongly believes in making her own decisions and more strongly distrusts the influence of others.

Upon their first full rehearsal, Mrs. Grant, who has taken the role of Cottager's wife, has to stay at home with her husband and thus Fanny is asked to read the part in her place: "as they all persevered—as Edmund repeated his wish, and with a look of even fond dependence on her good nature, she must yield" (MP, p. 201). But at that moment Sir Thomas returns home, and all plans for the play are scuttled. Explaining to his father what happened, Edmund affirms that "Fanny is the only one who has judged rightly throughout, who has been consistent" (MP, p. 219). Henry Crawford leaves despite Maria's hope that he would propose and save her from Mr. Rushworth. Sir Thomas cannot help noticing Mr. Rushworth's ignorance and Maria's dislike, and asks Maria if she really wants to marry him. Maria, furious at Henry's exit, wants to go through with it. The marriage takes place and Julia goes to live with the new couple, leaving Fanny the only young woman at Mansfield.

Fanny, who is previously not "out," now goes through the rituals accompanying the transition to womanhood. For the first time, she herself is invited to the parsonage by Mrs. Grant for dinner (before she dined out only along with her aunt and uncle), and Henry Crawford unexpectedly shows up. Challenged by Fanny's coldness, Henry tells his sister Mary that he aims to make Fanny fall in love with him while he is visiting for just a fortnight: "I will not do her any harm, dear little soul! I only want her to look kindly on me, to give me smiles as well as blushes" (MP, p. 269). But when her brother William receives leave from the navy and visits Mansfield, Henry is struck by Fanny's warmth toward her brother and decides to extend his stay. Sir Thomas notices Henry's interest in Fanny and organizes a ball for her; when Fanny discovers that she is to open the ball, "She could hardly believe it. To be placed above so many elegant young women! The distinction was too great. It was treating her like her cousins!" (MP, p. 320).

Soon afterward Henry proposes to Fanny. By working through his uncle Admiral Crawford, Henry has secured William's promotion to Second Lieutenant and thus appeals to Fanny's gratitude. When Fanny refuses, Henry appeals to Sir Thomas, who in turn lectures Fanny using arguments often made against rational choice theory: social embeddedness ("His sister, moreover, is your intimate friend, and he has been doing *that* for your brother, which I should suppose would have been almost sufficient recommendation to you"), undefined preferences ("I am half inclined to think, Fanny, that you do not quite know your own feelings"), excessive individuality ("The advantage or disadvantage of your family— of your parents—your brothers and sisters—never seems to have had a moment's share in your thoughts on this occasion. . . . You think only of yourself"), regret avoidance ("you may live eighteen years longer in the world, without being addressed by a man of half Mr. Crawford's estate, or a tenth part of his merits"), and duty ("had either of my daughters, on receiving a proposal of marriage at any time, which might carry with it only *half* the eligibility of *this*, immediately and peremptorily, and without paying my opinion or my regard the compliment of any consultation, put a decided negative on it. . . . I should have thought it a gross violation of duty and respect") (MP, pp. 364–68). Lady Bertram also appeals to duty ("it is every young woman's duty to accept such a very unexceptionable offer as this") and Mary Crawford appeals to social distinction ("were I to attempt to tell you of all the women whom I have known to be in love with him, I should never have done. . . . I am sure it is not in woman's nature to refuse such a triumph") (MP, pp. 384, 417–19). Against all these other models of human behavior, Fanny maintains that it is simply her choice: "I—I cannot like him, sir, well enough to marry him" (MP, p. 364).

Why don't people think of Fanny as making a choice? Refusing Henry would require a personhood and preferences that no one thinks she has. Fanny has never been allowed to choose in the past; why should she start now? Fanny's seeming inability to make independent choices is actually what Henry falls in love with, when he sees Fanny helping Lady Bertram "with such unpretending gentleness, so much as if it were a matter of course that she was not to have a moment at her own command" (MP, p. 343). Henry has preferences enough for the both of them: "He had vanity, which strongly inclined him, in the first place, to think she did love him, though she might not know it herself," and is "determined . . . to have the glory, as well as the felicity, of forcing her to love him" (MP, p. 376). Henry's sister Mary thinks not of Fanny's preferences but of her disposition and heart: "The gentleness and gratitude of her disposition would secure her all your own immediately.... [A]sk her to love you, and she will never have the heart to refuse" (MP, p. 340). Even Edmund, a formerly staunch advocate for Fanny's right to choose, presumes to know how Fanny feels: "I cannot suppose that you have not the *wish* to love him—the natural wish of gratitude. You must have some feeling of that sort. You must be sorry for your own indifference" (MP, p. 402). When Fanny cries, "Oh! never, never, never; he never will succeed with me," Edmund thinks he knows Fanny better than she knows herself and scolds, "Never, Fanny, so very determined and positive! This is not like yourself, your rational self" (MP, p. 402).

When Fanny tries to talk about her own preferences, no one listens. When Sir Thomas touches on the topic of whether Fanny's affections lay elsewhere, Fanny's "face was like scarlet. That, however, in so modest a girl, might be very compatible with innocence; and chusing at least to appear satisfied, he quickly added, 'No, no, I know *that* is quite out of the question—quite impossible'" (MP, p. 365). She dare not tell anyone about her love for Edmund. She cannot tell Sir Thomas that her contempt for Henry stems from his reprehensibly superficial overtures toward Maria and Julia, in which they eagerly participated. When Fanny does try to tell him that she simply does not like Henry, it doesn't work at all: "She had hoped that to a man like her uncle, so discerning, so honourable, so good, the simple acknowledgment of settled *dislike* on her side, would have been sufficient. To her infinite grief she found it was not" (MP, pp. 366–67). When Fanny tries to bring up Henry's treatment of Maria and Julia, both Mary Crawford and Edmund are willing to excuse Henry at the sisters' expense. Mary owns, "He has now and then been a sad flirt, and cared very little for the havock he might be making in young ladies' affections. I have often scolded him for it, but it is his only fault; and there is this to be said, that very few young ladies have any affections worth caring for" (MP, p. 419). Edmund says that his sisters might "be more desirous of being admired by Crawford, and might shew that desire rather more unguardedly than was perfectly prudent.... [W]ith such encouragement, a man like Crawford, lively, and it may be a little unthinking, might be led on" (MP, p. 405).

Sir Thomas worries about Henry's patience running out and decides to speed up Fanny's reconsideration by sending her back to visit her parents in Portsmouth. Sure enough, her family's noisy and crowded living conditions, with three noisy young boys and two daughters still at home, heighten the salience of Mansfield's comforts and the similar life that Henry's wealth would provide.

In Portsmouth, Fanny's fourteen-year-old sister Susan seems aware that the house is not managed well, and Fanny endeavors to help her.

"Susan, she found, looked up to her and wished for her good opinion; and new as any thing like an office of authority was to Fanny, new as it was to imagine herself capable of guiding or informing any one, she did resolve to give occasional hints to Susan...which her own more favoured education had fixed in her" (MP, p. 459). In chaper 2, Susan has been quarreling with Betsey, who is five years old, about a silver knife that their sister Mary gave to Susan before her death; it is Susan's but Betsey does not want to give it up. Fanny purchases a new knife for Betsey, who prefers the newer object, as Fanny predicts, and thus the original knife returns to Susan. "The deed thoroughly answered; a source of domestic altercation was entirely done away, and it was the means of opening Susan's heart to her, and giving her something more to love and be interested in" (MP, pp. 459–60). This is Fanny's first successful nontrivial strategic manipulation, another part of her favored education at Mansfield; strategic thinking both initiates the loving bond between her and Susan and provides a shared interest.

In her new role as teacher, Fanny is surprised that Susan has much good sense, given that she "had no cousin Edmund to direct her thoughts or fix her principles" (MP, p. 460). She is to Susan what Edmund was to her; Fanny purchases a subscription to a circulating library and "longed to give her a share in her own first pleasures, and inspire a taste for the biography and poetry which she delighted in herself" (MP, p. 461). The one positive aspect of marrying Henry Crawford that Fanny is willing to consider is the possibility of Susan living with them and thereby improving, as Fanny is worried about what will become of Susan: "That a girl so capable of being made everything good, should be left in such hands, distressed her more and more. Were *she* likely to have a home to invite her to, what a blessing it would be!—And had it been possible for her to return Mr. Crawford's regard, the probability of his being very far from objecting to such a measure, would have been the greatest increase of all her own comforts" (MP, p. 486). Developing the minds of young women is so important to Fanny that she is willing to contemplate what she desires least.

While Fanny remains in Portsmouth, Tom Bertram falls ill, and declines enough so that Mary Crawford begins to consider that Edmund, next in line, would make a more worthy husband as heir to Mansfield Park than as a clergyman. The greater scandal, however, is that Henry Crawford and Maria Bertram, now Mrs. Rushworth, have run off together. Edmund is stunned that Mary Crawford sees her brother's action "only as folly, and that folly stamped only by exposure....Oh! Fanny, it was the detection, not the offence which she reprobated" (MP, p. 526). Edmund is sorrowful but Fanny is happy that at long last, "Edmund was no longer the dupe of Miss Crawford" (MP, p. 533).

In his infatuation with Mary Crawford, Edmund forgets one of the first precepts of strategic thinking: a person's motivations are often different from what you think they are or should be. After his final conversation with Mary, Edmund concludes that "all this together most grievously convinced me that I had never understood her before, and that, as far as related to mind, it had been the creature of my own imagination, not Miss Crawford, that I had been too apt to dwell on for many months past" (MP, p. 530). His family's reputation ruined, Sir Thomas also reflects upon the same mistake: "Bitterly did he deplore a deficiency which now he could scarcely comprehend to have been possible.... [H]e had brought up his daughters, without ... being acquainted with their character and temper" (MP, p. 536). At least Mr. Rushworth, when marrying Maria, does not make this mistake: Maria "had despised him, and loved another—and he had been very much aware that it was so" (MP, p. 537).

Similarly, Mary Crawford does not understand Edmund's motivations, the seriousness of his commitment to becoming a clergyman. Mary's first mistake, making fun of the clergy before knowing Edmund's plan to join, was an error of information ("Fanny pitied her. 'How distressed she will be at what she said just now,' passed across her mind" [MP, p. 104]). But Mary's persistence in believing him "fit for something better," even after talking with him at length, is an error of understanding (MP, p. 109). To Mary, Edmund's motivations are what she thinks and hopes they are, not what they truly are. If Mary had understood Edmund's commitment, she would not have tried to intimate that her acceptance of his proposal was dependent on his career choice, knowing that he would find this intimation particularly repugnant. Henry Crawford's mistake when flirting with the new Mrs. Rushworth was underestimating the strength of Maria's feelings: he had only meant to feed his own vanity but "he had put himself in the power of feelings on her side more strong than he had supposed.—She loved him; there was no withdrawing attentions, avowedly dear to her" (MP, p. 541).

Fanny is basically right about everyone else, and everyone else is wrong about Fanny. This is because Fanny is quiet, a good listener; appearing passive has its advantages. During play rehearsals, "Fanny, being always a very courteous listener, and often the only listener at hand, came in for the complaints and the distresses of most of them" (MP, p. 192). When Fanny returns to Mansfield Park after Maria has run off with Henry Crawford, she supports Lady Bertram simply by listening: "To talk over the dreadful business with Fanny, talk and lament, was all Lady Bertram's consolation. To be listened to and borne with, and hear the voice of kindness and sympathy in return, was every thing that could be done for her" (MP, p. 519). Listening skills can even make someone fall in love with you; Henry Crawford declares, "I could so wholly and absolutely confide in her . . . and *that* is what I want" (MP, p. 341).

It is true that Fanny never seems to do anything to advance her goal of securing Edmund's heart. But this underestimates the power of active listening, of active agreeing. From early on, Fanny, a "kind, kind listener," strongly agrees with Edmund about the flaws in Mary Crawford's character, and strongly agrees that they are the result of poor upbringing, "[t]he effect of education" (MP, p. 312). When Mary fails to condemn her brother for running off with Maria, and especially after Fanny tells Edmund that Mary's interest in him was increased by Tom's illness, Fanny and Edmund "thought exactly the same; and they were also quite agreed in their opinion of the lasting effect, the indelible impression, which such a disappointment must make on his mind" (MP, pp. 531–32). Fanny strongly agrees that it is impossible that he will ever meet another woman who can ease his mind and that "Fanny's friendship was all that he had to cling to" (MP, p. 532). The situation is not that different from Lucy Steele's taking in of Robert Ferrars by listening at length, in repeated visits, to his argument that she cannot marry Edward Ferrars in *Sense and Sensibility*, or Charlotte Lucas's receiving Mr. Collins's proposal by tending to his rejection by Elizabeth Bennet in *Pride and Prejudice* (Charlotte's "civility in listening to him, was a seasonable relief to them all" [PP, p. 129]).

Fanny is more than happy to provide talk therapy to Edmund in soothing locations, and "[a]fter wandering about and sitting under trees with Fanny all the summer evenings, he had so well talked his mind into submission, as to be very tolerably cheerful again" (MP, p. 535). Edmund's love for Fanny began when Fanny was ten, "founded on the most endearing claims of innocence and helplessness." If helplessness provided the initial attraction, Edmund's regard is "completed by every recommendation of growing worth," now that Fanny has learned and developed so much. "And being always with her, and always talking confidentially . . . those soft light eyes could not be very long in obtaining the pre-eminence" (MP, p. 544).

Meek Fanny is actually the most hard-core of Austen's heroines; only Fanny makes a decision in the face of everyone's active opposition, without a single supporter. Had Fanny accepted Henry Crawford's proposal, "Henry would have been too happy and too busy to want any other object" and would not have run off with the married Maria, according to Mary Crawford (MP, p. 527). This counterfactual is not farfetched, as even Fanny is not completely unpersuadable; had Henry persisted, "Fanny must have been his reward—and a reward very voluntarily bestowed—within a reasonable period from Edmund's marrying Mary" (MP, p. 540). The Bertram family reputation would

have remained intact and Edmund might have been tolerably happy
with Mary. It is even probable that Fanny would have been happy as
Mrs. Crawford, by focusing her energies on improving Susan. By refusing
Henry, Fanny in the end gains her first object, but at the great expense,
even ruin, of the Crawfords and the Bertrams, Edmund perhaps excepted.
So much for Fanny being self-sacrificing. As Fanny observes, "We have
all a better guide in ourselves, if we would attend to it, than any other
person can be" (MP, p. 478).

EMMA

Emma Woodhouse at twenty is already the mistress of her own household at Hartfield, in the village of Highbury, where she lives with her
father, Mr. Woodhouse. Her female superiors have departed: her mother
died years ago, and her elder sister Isabella and her governess Miss Taylor
have married and moved out. Emma has strategic skills and the social
position to use them. But believing too much in one's own strategic
ability has its pitfalls, which Emma learns the hard way. "The real evils,
indeed, of Emma's situation were the power of having rather too much
her own way, and a disposition to think a little too well of herself.... The
danger, however, was at present so unperceived" (E, pp. 3–4). Just as
Brer Rabbit thinks the tar baby rude, and just as Marianne Dashwood,
hoping for a letter from Willoughby in *Sense and Sensibility*, thinks
Mrs. Jennings cruel for handing her a letter from her mother, Emma's
overconfidence in her ability to figure out others' motives leads to a
solipsism just as bad as thinking that they have no independent thoughts
at all. Pride in one's strategic skills can lead to their overextension for the
purpose of impressing others and also leaves one open to manipulation
through flattery. Too much attention to the preferences of others can
keep you from considering the preferences that are most important, your
own. *Emma* is not a first course in strategic thinking but a corrective for
those impressed by their own abilities.
    Emma takes pride in her strategic skill at the very start, claiming
credit for matching Miss Taylor with her new husband, Mr. Weston.
Mr. George Knightley, the brother of Isabella's husband, Mr. John
Knightley, disagrees, saying that Emma may have favored their match,
but "[s]uccess supposes endeavour.... [Y]ou made a lucky guess; and
that is all that can be said" (E, p. 11). But Emma counters, "If
I had not promoted Mr. Weston's visits here, and given many little encouragements, and smoothed many little matters, it might not
have come to any thing after all" (E, pp. 11–12). Mr. Woodhouse,
"understanding but in part," is charmingly clueless in strategic matters

and Emma effortlessly directs him, taking his mind off the sadness of Miss Taylor's departure by saying that it was he, not Emma herself, who arranged for the daughter of their servant James to be Miss Taylor's housemaid, so that their households would frequently communicate (E, p. 12). In the minor matter of the placement of James's daughter, Emma generously grants her father credit for what most likely originated with her own suggestion. However, for Miss Taylor's marriage itself, Emma eagerly claims responsibility. To answer Mr. Knightley's challenge to her efficacy, Emma declares that her next match will be for the benefit of Mr. Elton, the vicar of Highbury; "it would be a shame to have him single any longer" (E, p. 12).

Emma comes upon seventeen-year-old Harriet Smith, who has been brought up at the nearby boarding school for girls and has no observable parents or other relatives. Emma takes up Harriet's improvement as her personal project (acting like Henry Crawford toward Fanny Price in *Mansfield Park*): "*She* would notice her; she would improve her; she would detach her from her bad acquaintance, and introduce her into good society; she would form her opinions and her manners" (E, p. 23). Most worrisome is Harriet's closeness with the Martin family, who rents a farm from Mr. Knightley and includes two sisters as well as the unmarried brother Mr. Robert Martin. Emma thinks the higher-status Mr. Elton would well displace the farmer Mr. Martin in Harriet's consideration. Emma's concern is not just whether her plan will work but also whether it gives distinction to herself as a strategist: "She feared it was what every body else must think of and predict. It was not likely, however, that any body should have equalled her in the date of the plan, as it had entered her brain during the very first evening of Harriet's coming to Hartfield" (E, pp. 34–35). Even if others may have had the same idea, Emma thought of it first. Emma arranges to paint a watercolor of Harriet, with Mr. Elton present to provide encouragement. Mr. Elton volunteers to take the finished portrait to London to have it framed.

When Harriet receives Mr. Martin's marriage proposal by letter, she runs to Emma for advice. Emma dispassionately states, "I shall not give you any advice, Harriet.... This is a point which you must settle with your own feelings," only to add, "It is not a state to be safely entered into with doubtful feelings, with half a heart" (E, pp. 54, 55). Thus browbeaten, Harriet decides to refuse the proposal and Emma vehemently agrees, saying that she is relieved, because as Mrs. Martin, Harriet would have been "confined to the society of the illiterate and vulgar" (E, p. 56). To cheer Harriet up, Emma constructs a fabulous scenario of Mr. Elton's intimacy with Harriet's portrait: "At this moment, perhaps, Mr. Elton is shewing your picture to his mother and sisters, telling how much more beautiful is the original.... It is his companion all this evening, his solace, his delight. It opens his designs to his family, it introduces you among them. . . . How cheerful, how animated, how suspicious, how busy their imaginations all are!" (E, p. 59). For Emma, part of what makes this vision so satisfactory is that she and Harriet know something that others only suspect.

Mr. Knightley visits Emma with news: "I have good reason to believe your little friend will soon hear of something to her advantage" (E, p. 62). Emma, "smiling to herself," says that she already knows of Mr. Martin's proposal and that it was refused (E, p. 63). A stunned Mr. Knightley accuses Emma of persuading Harriet, and Emma replies that even if she had, she would have been justified in doing so. Emma's argument is not based on Harriet's right to make her own choice ("as to the refusal itself, I will not pretend to say that I might not influence her a little" [E, p. 68]). Emma argues that Harriet has alternatives ("is she, at seventeen, just entering into life, just beginning to be known, to be wondered at because she does not accept the first offer she receives? No—pray let her have time to look about her" [E, p. 67]), but her main argument is based on social status ("The sphere in which she moves is much above his.—It would be a degradation. . . . [S]he associates with gentlemen's daughters" [E, pp. 65–66]), a status that Emma herself has worked hard to confer. What Harriet actually wants does not enter into the discussion. Mr. Knightley warns that Mr. Elton would never choose Harriet: "He knows the value of a good income as well as anybody" (E, p. 70).

Mr. Elton's wooing continues, and just as Elizabeth Bennet in *Pride and Prejudice* understands Mr. Darcy's invitation to dance as showing contempt instead of interest, Emma's strategic sophistication allows her to explain away the obvious in favor of her own preconceptions. Emma asks Mr. Elton to add to Harriet's collection of riddles, and when he pushes the riddle he composed toward Emma instead of Harriet, Emma understands this as showing the boldness of his love for Harriet, because if he had intended it to be secret, he would have given it to Harriet when Emma was not present. When Mr. Elton uses the term "ready wit" in his riddle (E, p. 119), Emma concludes that only a man very much in love would use the term to refer to not-so-quick Harriet, and wishes that she could show Mr. Knightley the riddle so he would have to admit that he is mistaken about Mr. Elton. When Harriet and Mr. Elton are invited by the Westons to join them in a dinner with the Woodhouses and Knightleys at their home at Randalls, Harriet is too ill to join, and Emma provides an out for Mr. Elton by saying that he looks slightly ill and should stay home also. Mr. Elton nevertheless joins the party, and Emma concludes that he must be "very much in love with Harriet; but still, he cannot refuse an invitation, he must dine out wherever he is asked. What a strange thing love is!" (E, p. 119). At the party, Mr. Elton treats Emma with "extreme solicitude.... It did appear—there was no concealing it—exactly like the pretence of being in love with her, instead of Harriet" (E, p. 135).

Returning from the party in a carriage alone together, Emma is astonished and repulsed by Mr. Elton's proposal. Emma accuses Mr. Elton of having transferred his affections from Harriet, but Mr. Elton is surprised that Emma even mentions Harriet. Mr. Elton, "very much resolved on being seriously accepted as soon as possible," is presumptuous in turn, saying to Emma, "I am sure you have seen and understood me," and claiming even that Emma had encouraged him (E, pp. 140, 142). Instead of Emma providing cover for Mr. Elton's approach toward Harriet, Harriet provided cover for Mr. Elton's approach toward Emma.

Reflecting on her folly, Emma blames Mr. Elton's low social status: "Perhaps it was not fair to expect him to feel how very much he was her inferior in talent, and all the elegancies of mind. The very want of such equality might prevent his perception of it" (E, p. 147). Only reluctantly does she admit to herself that their misperceptions were mutual: "Emma was obliged in common honesty to stop and admit that her own behaviour to him had been so complaisant and obliging.... If *she* had so misinterpreted his feelings, she had little right to wonder that *he*, with self-interest to blind him, should have mistaken her's" (E, pp. 147–48). Emma acknowledges that Mr. Knightley was right about Mr. Elton and she was wrong; even worse, Mr. Knightley's brother John had earlier warned her of Mr. Elton's apparent interest but Emma had laughed it off, "amusing herself in the consideration of the blunders which often arise from a partial knowledge of circumstances, of the mistakes which people of high pretensions to judgment are for ever falling into" (E, pp. 120–21). Even when she has to tell Harriet that Mr. Elton never loved her, Emma still thinks of Harriet's tearful reaction in terms of social distinction, superior versus inferior: "her grief was so truly artless, that no dignity could have made it more respectable in Emma's eyes—and she listened to her and tried to console her with all her heart and understanding—really for the time convinced that Harriet was the superior creature of the two" (E, p. 153).

Emma would have figured out eventually that she, not Harriet, was Mr. Elton's target; had Emma and Mr. Elton not been unfortunately placed alone in a carriage together, their misperceptions would not have been so grossly revealed. But why did Emma's misperceptions persist for as long as they did? Her strategicness helps explain away discrepancies, but on top of this her sense of entitlement and superior social status allows her to conclude that when another person does not do what she thinks he should do, there is something wrong with him rather than something wrong with her own perception. When Mr. Elton eagerly agrees with Emma's suggestion of creating a portrait of Harriet, saying that he delights in seeing "specimens of your landscapes and flowers," Emma says to herself, "You know nothing of drawing. Don't pretend to be in raptures about mine. Keep your raptures for Harriet's face" (E, pp. 44, 45). When Mr. Elton entreats Emma not to visit the ill Harriet and protect her own health, instead of concluding that maybe he does not love Harriet, Emma sees in Mr. Elton "an inconstancy, if real, the most contemptible and abominable! and she had difficulty in behaving with temper" (E, p. 135). In this respect, Emma is just a less unpleasant version of Lady Catherine, who when stooping to visit the lower-status Elizabeth Bennet in *Pride and Prejudice*, presumes that Elizabeth already knows the reason for her visit.

A fixation with status and gaining distinction for her own strategic ability is what makes Emma embark on matching Harriet and Mr. Elton in the first place. When Emma contrives for Harriet and herself to visit the vicarage, and successfully retreats so that Harriet and Mr. Elton are alone together, "Emma felt the glory of having schemed successfully" (E, p. 96). It is not enough for Mr. Elton simply to love Harriet; Harriet must also feel proud of the strategizing that went into it: "I congratulate you, my dear Harriet, with all my heart. This is an attachment which a woman may well feel pride in creating" (E, p. 78). The achievement of Harriet's match is measured in public comparison with other marriages ("Your marrying will be equal to the match at Randalls" [E, p. 79]), not in the couple's happiness. Emma's goal from the start is to demonstrate her strategic prowess to Mr. Knightley and others, and getting Harriet married is just a vehicle toward that end. Emma took special pride in matching Miss Taylor with Mr. Weston because "so many people said Mr. Weston would never marry again" (E, p. 10). Thus for her strategic prowess to attain still greater distinction, Emma needs an even more unlikely couple. Emma is strategically sophisticated but obsessed with what her strategic actions "mean" in terms of social distinction. In comparison, Mr. Knightley has substantial strategic skills of his own but plays them down: "I do not pretend to Emma's genius for foretelling and guessing" (E, p. 38). After her failure, seeing Harriet's tears, Emma cannot abandon her strategic skills but tries to be less demonstrative: "It was rather too late in the day to set about being simple-minded and ignorant; but she left her with every previous resolution confirmed of being humble and discreet" (E, p. 153).

Humbled in manipulation, Emma falls back on flaunting her skills of detection. Of course the most important preference to detect is who is in love with whom. Jane Fairfax, who is the same age as Emma, arrives to stay with her aunt Miss Bates and her grandmother Mrs. Bates. Since her own parents died early, Jane was raised by Colonel and Mrs. Campbell, together with their own daughter Miss Campbell, who has recently married Mr. Dixon and moved with him to Ireland. The Campbells are visiting Ireland but Jane decides to visit Highbury instead, and Emma quickly surmises that Jane is thereby avoiding Mr. Dixon because of a secret attachment between them. For Emma, this conclusion is not at all farfetched, since Jane Fairfax is reportedly far more beautiful than the former Miss Campbell, Mr. Dixon openly prefers Jane's singing to his wife's, and Mr. Dixon kept Jane from falling off a boat.

With little wealth or family support, Jane Fairfax is preparing for life as a governess. Emma dislikes Jane because she is so reserved, providing no real information about Frank Churchill, for example. Frank Churchill is the son of Mr. Weston from his first marriage to Miss Churchill; after her early death he was raised by her brother's family and later adopted their surname. It is expected that Frank Churchill must soon visit to pay his respects to his father's new wife, and Emma thinks him marked off for herself: "if she *were* to marry, he was the very person to suit her in age, character and condition. He seemed by this connection between the families, quite to belong to her" (E, p. 128).

When Frank Churchill finally obtains leave from his needy and illness-prone aunt, Mrs. Churchill, Emma observes his warm and open compliments; he is indeed very good at "knowing how to please" (E, p. 206). She finds it odd that he goes all the way to London just to get a haircut, but thinking to herself defends him against what she imagines to be Mr. Knightley's reproach: "Mr. Knightley, he is *not* a trifling, silly young man. If he were, he would have done this differently. He would either have gloried in the achievement, or been ashamed of it. There would have been either the ostentation of a coxcomb, or the evasions of a mind too weak to defend its own vanities" (E, p. 229). For Emma, who is preoccupied with presentation, what indicates Frank Churchill's character is not the decision itself but how he publicly presents it. Later, Emma is busy "fancying what the observations of all those might be, who were now seeing them together for the first time" (E, p. 230). A pianoforte mysteriously arrives at the Bates residence for Jane Fairfax, and when Emma suggests that it must have come not from her guardian, Colonel Campbell, but from Mr. Dixon, Frank Churchill agrees, "I can see it in no other light than as an offering of love" (E, p. 236).

Frank Churchill quickly becomes Emma's ally in investigating and otherwise tormenting Jane Fairfax. Miss Bates invites them to give their opinion on the new pianoforte. While Jane Fairfax sits at the instrument, Frank Churchill asks her, "How much your friends in Ireland must be enjoying your pleasure on this occasion, Miss Fairfax. I dare say they often think of you, and wonder which will be the day, the precise day of the instrument's coming to hand. Do you imagine Colonel

Campbell knows the business to be going forward just at this time?" Emma "wished he would be less pointed, yet could not help being amused," and whispers, "You speak too plain. She must understand you" (E, pp. 260, 262). Emma begins to accept that her strategic companion Frank Churchill is in love with her, especially after he begins an awkward confession, only to be interrupted by his father walking in. Emma inquires as to whether she is in love with Frank Churchill, but decides otherwise. She reminds herself not to encourage him and wonders whether Harriet might be a good consolation prize.

Mrs. Weston suggests to Emma that Mr. Knightley might be in love with Jane Fairfax, and even that he bought the pianoforte himself, given his obvious admiration of her performances. Mr. Knightley had also sent his own carriage to transport Jane and Miss Bates on a cold evening and had even sent to the Bates residence his complete supply of the particular apples that Jane Fairfax enjoys. Emma deems this prospect a "very shameful and degrading connexion" and sets out to destroy any idea of it (E, p. 243). Emma notes that Mr. Knightley does not care for the upcoming ball, which Jane will attend, and successfully contrives his explicit statement that he will never ask Jane to marry him. Mr. Knightley tells Emma that she is "miserably behindhand" and that he has already disabused Mr. Cole of any notion that he might marry Jane: "Mr. Cole gave me a hint of it six weeks ago.... I told him he was mistaken; he asked my pardon and said no more. Cole does not want to be wiser or wittier than his neighbours" (E, pp. 310–11).

At the ball, Mr. Elton pointedly refuses Mrs. Weston's request that he might dance with Harriet, leaving her the only young woman without a partner. But Mr. Knightley, who does not usually dance, steps in and asks Harriet, delighting Emma. Harriet is rescued again the next day, this time by Frank Churchill, who chases away some gypsy children begging Harriet for money. Emma observes, "[I]t had happened to the very person, and at the very hour, when the other very person was chancing to pass by to rescue her!—It certainly was very extraordinary!...It was not possible that the occurrence should not be strongly recommending each to the other" (E, pp. 362–63). Emma gets Harriet to express her admiration, without naming names, of the gentleman's "infinite superiority" and her gratitude for being rescued: "[W]hen I saw him coming—his noble look—and my wretchedness before. Such a change! ...From perfect misery to perfect happiness!" (E, p. 370).

Mr. Knightley has been watching Frank Churchill, and when he sees him look expressively at Jane Fairfax, he senses a private connection between them. When he tries to warn Emma, she is too embarrassed to reveal that she and Frank Churchill have their own joint project of suspecting Jane Fairfax's illicit connection with Mr. Dixon, and she emphatically declares that Jane Fairfax and Frank Churchill "are as far from any attachment or admiration for one another, as any two beings in the world can be." Mr. Knightley is "staggered" by Emma's confidence (E, p. 380).

Emma and Frank Churchill try to enliven a dull conversation with a game in which everyone has to say one clever thing or three dull things. Miss Bates volunteers that she can easily come up with three dull things, and Emma cannot help chiming in that Miss Bates might have difficulty limiting herself to only three. Later Mr. Knightley reprimands Emma for this impudence. Emma defends herself by saying, "[W]hat is good and what is ridiculous are most unfortunately blended in her"; Mr. Knightley agrees but explains that it is not the literal meaning of Emma's remark but its social context that makes it particularly offensive: "She is poor; she has sunk from the comforts she was born to; and, if she live to old age, must probably sink more.... You, whom she had known from an infant, whom she had seen grow up from a period when her notice was an honour, to have you now, in thoughtless spirits, and the pride of the moment, laugh at her, humble her—and before her niece, too" (E, p. 408). Emma reflects upon her cruelty and uncharacteristically cries on the way home.

After delaying for months, Jane Fairfax finally submits to taking a position as a governess, at the arrangement of Mrs. Elton, Mr. Elton's new wife. Luckily, Frank Churchill's aunt Mrs. Churchill dies, allowing him to gain permission to marry from his more pliable uncle and finally reveal his secret engagement with Jane Fairfax. Emma was correct in thinking the piano a love offering, but it was from Frank Churchill himself rather than Mr. Dixon; Frank Churchill had arranged the purchase while he was in London getting his haircut. Emma's error was thinking the haircut a mark of vanity instead of simply a cover for a specific instrumental action. Mr. Knightley's suspicions were correct; Emma dismissed them because of her own presumed strategic partnership with Frank Churchill. Frank Churchill duped Emma by playing to her self-regard as a strategist; excessive pride in one's strategic ability invites flattery and thereby opens up another way to be manipulated. As W. C. Fields says, "you can't cheat an honest man" (Marshall 1939). After all, Emma's suspicion of Mr. Dixon was initially her own idea, and Frank Churchill merely went along with it. He had tried to tell Emma of his secret, and this, not a marriage proposal, was the awkward confession which was interrupted by his father. Mr. Knightley is correct in saying that Frank Churchill makes Emma's "talents conduce to the display of his own superiority," and Frank Churchill's vanity about his own strategic ability makes him also overextend and nearly blunder (E, p. 162). In open conversation he assumes that Mr. Perry's plan to set up a carriage is publicly known when it was actually discussed only among the Bates household (Mrs. Bates, Miss Bates, and Jane Fairfax); he thus almost reveals his secret communication with Jane Fairfax.

Emma worries about disheartening Harriet again, this time with the news of Frank Churchill's engagement, but Harriet is surprised that Emma even mentions him; the rescuer whom Harriet admires is Mr. Knightley. Emma realizes that Mr. Knightley has been paying greater attentions to Harriet recently, and a "mind like her's, once opening to suspicion, made rapid progress.... Why was it so much worse that Harriet should be in love with Mr. Knightley, than with Frank Churchill? Why was the evil so dreadfully increased by Harriet's having some hope of a return?" (E, p. 444). Only with a competitor's active threat does Emma realize that Mr. Knightley must marry herself. Immediately Emma knows what she must do to protect her own claim: "For her own advantage indeed, it was fit that the utmost extent of Harriet's hopes should be enquired into" (E, p. 445). Emma guides the conversation to get as much detail as possible about why Harriet thinks Mr. Knightley is interested in her. After Harriet leaves, Emma for the first time inquires deeply into her own feelings: "To understand, thoroughly understand her own heart, was the first endeavour.... Till now that she was threatened with its loss, Emma had never known how much of her happiness depended on being *first* with Mr. Knightley, first in interest and affection" (E, pp. 449–52).

Mr. Knightley is around thirty-seven and "one of the few people who could see faults in Emma Woodhouse, and the only one who ever told her of them.... [H]e had loved her, and watched over her from a girl, with an endeavour to improve her, and an anxiety for her doing right" (E, pp. 9, 452). But especially after his excoriation of her behavior toward Miss Bates, Emma worries that Harriet might have greater claim to his affection.

Mr. Knightley comes to see Emma, thinking that she must be disappointed about Frank Churchill's engagement. Emma assures him that she never had any real interest in Frank Churchill, and Mr. Knightley confesses, "In one respect he is the object of my envy" (E, p. 467). Even after chastising herself for her presumption ("With insufferable vanity had she believed herself in the secret of every body's feelings; with unpardonable arrogance proposed to arrange everybody's destiny" [E, p. 449]), she still tries to manage the situation by changing the subject. Emma does not allow Mr. Knightley to simply make his statement; Emma thinks he is talking about Harriet and asks him not to speak. But Emma crucially reconsiders, thinking that she must allow Mr. Knightley to confide in her: "cost her what it would, she would listen. She might assist his resolution, or reconcile him to it; she might give just praise to

Harriet, or, by representing to him his own independence, relieve him from that state of indecision" (E, p. 468). She tells Mr. Knightley that she will continue walking with him and that he should speak openly to her as a friend.

Mr. Knightley submits, "You hear nothing but truth from me.—I have blamed you, and lectured you, and you have borne it as no other woman in England would have borne it.—Bear with the truths I would tell you now, dearest Emma, as well as you have borne with them.... God knows, I have been a very indifferent lover.—But you understand me" (E, p. 469). Emma finally understands Mr. Knightley's regard. Emma's most important strategic move, continuing their walk, taking another turn around the grounds and allowing Mr. Knightley to speak, does not need to be crowed about or gloried in; all that matters is that it works. In fact, Emma feels its inelegance: "her proposal of taking another turn, her renewing the conversation which she had just put an end to, might be a little extraordinary!—She felt its inconsistency; but Mr. Knightley was so obliging as to put up with it, and seek no farther explanation" (E, p. 470). Mr. Knightley's lectures and program of instruction for Emma over the years, and Emma's willingness to be instructed, are the basis for his love for her. Thus his lecture about Emma's treatment of Miss Bates is an expression of care and affection, not dislike. Emma's commitment to being Mr. Knightley's strategic partner, to help him make a decision and increase his resolve, even to the benefit of her rival, motivates her decision to continue their walk and is therefore the basis for her success in the end. She succeeds this one time, not by correctly inferring another's motivations or by cleverly manipulating him, but by simply allowing that person the autonomy to speak for himself.

Mr. Knightley's jealousy of Frank Churchill "probably enlightened him as to" his love for Emma, and thus both his and Emma's affection were catalyzed by rivals (E, p. 471). Harriet's strategic skills have developed: her perception of Mr. Elton's affection was dominated by Emma's interpretations and was completely wrong, but she is able to detect Mr. Knightley's interest all by herself (initially skeptical of Emma's high opinion of Harriet, Mr. Knightley had taken efforts to get to know her). Mr. Robert Martin proposes to Harriet again, and this time Harriet accepts without any advice necessary, outgrowing Emma's tutelage just as Anne Elliot in *Persuasion* outgrew Lady Russell's.

Emma is the least constrained of all Austen's heroines, but is hardly the most independent; she craves praise and validation and thinks constantly about how her strategic actions will be reviewed by others, particularly by her tutor, Mr. Knightley. This is understandable since Emma does not have anyone else around who is expert enough to appreciate her skills. Elizabeth Bennet has her father in *Pride and Prejudice*, Elinor and

Marianne Dashwood have their mother in *Sense and Sensibility*, Anne Elliot in *Persuasion* is too mature to care about showing her skills off, and Catherine Morland in *Northanger Abbey* and Fanny Price in *Mansfield Park* are too young to know that they can. Mr. Knightley's criticisms are unsparing (for example, "Better be without sense, than misapply it as you do" [E, p. 67]), but at least he understands. Emma's father is hopeless strategically and her sister Isabella is "slow and diffident" (E, p. 37). Her former governess Mrs. Weston is not great either. Mrs. Weston should have been able to detect the potential of Emma and Mr. Knightley (Mr. Knightley had agreed with her statement that Emma was "loveliness itself" [E, p. 39], and she had witnessed Emma bristle when she spoke of a match between Mr. Knightley and Jane Fairfax), but she does nothing, worrying instead about disappointing Emma with the news of Frank Churchill's engagement. Emma needs a colleague, a co-conspirator, and what makes Emma excited about the arrival of Frank Churchill is her anticipation that his strategic acumen might appreciate hers. Emma's match with Mr. Knightley finally gives her that true strategic partner.

Austen's Foundations of Game Theory

AUSTEN CAREFULLY establishes game theory's core concepts: choice (a person takes an action because she chooses to do so), preferences (a person chooses the action with the highest payoff), and strategic thinking (before taking an action, a person thinks about how others will act). A person's preferences are best revealed by her choices, and strategic thinking has several names, including "penetration." Austen illustrates (the lack of) strategic thinking through her strategic sophomores, characters who think they are skilled but are not. Her strategically skilled characters know how to detect a person's preferences by observing their eyes. In this and the next six chapters, I take Austen's six novels together as one body of work.

CHOICE

For Austen, choice is a central concern, even obsession. The single most important choice is a woman's choice of whether and whom to marry, and Austen's heroines adamantly defend this choice against any presumption otherwise. After Edward Ferrars is disowned by his family for his secret engagement with Lucy Steele, John Dashwood tells his sister Elinor Dashwood that the Ferrars family now plans for the wealthy Miss Morton to marry Edward's brother Robert instead. Elinor could not care less about Miss Morton, but defends the principle: "The lady, I suppose, has no choice in the affair" (SS, p. 336). Edmund Bertram tells Fanny Price that Henry Crawford's sisters are surprised by her refusal of Henry's proposal, but Fanny submits, "Let him have all the perfections in the world, I think it ought not to be set down as certain, that a man must be acceptable to every woman he may happen to like himself" (MP, p. 408). As mentioned in chapter 1, after Harriet Smith refuses Mr. Robert Martin's proposal, Emma Woodhouse responds to Mr. Knightley's indignation by saying, "[I]t is always incomprehensible to a man that a woman should ever refuse an offer of marriage. A man always imagines a woman to be ready for anybody who asks her" (E, p. 64). When Lady Catherine commands Elizabeth Bennet to promise never to become engaged to Mr. Darcy, Elizabeth defends her right to choose: "I am only resolved to act in that manner, which will, in

my own opinion, constitute my happiness, without reference to *you*, or
to any person so wholly unconnected with me" (PP, p. 396). According
to Johnson (1988, p. 84), "among Austen's contemporaries perhaps only
declared radicals would have a sympathetic character defiantly" make
such an undiluted statement. In contrast, the unheroic, self-centered
Mary Musgrove, Anne Elliot's sister, would side with Lady Catherine,
stating, "I do not think any young woman has a right to make a choice
that may be disagreeable and inconvenient to the *principal* part of her
family" (P, p. 82).

Thoughtful men are aware that women can make choices. Edmund
Bertram asks Mrs. Norris to stop pressuring Fanny to act in the play and
let Fanny "choose for herself as well as the rest of us" (MP, p. 172). Frank
Churchill states, "It is always the lady's right to decide on the degree
of acquaintance" (E, p. 216). Idiotic men are not aware, as Johnson
(1988, p. 36) observes: "Many Austenian men...cannot take 'no' for
an answer." John Thorpe does not know that dancing with a woman
requires asking her; he goes up to Catherine Morland and declares,
"Well, Miss Morland, I suppose you and I are to stand up and jig it
together again" (NA, p. 54).

Being able to make a choice is almost always a good thing for Austen;
it is "a great deal better to chuse than to be chosen" (E, p. 15). There
is power in being able to make a choice. Elizabeth Bennet remarks that
Mr. Darcy has "great pleasure in the power of choice. I do not know
any body who seems more to enjoy the power of doing what he likes
than Mr. Darcy" (PP, p. 205). Similarly, Emma believes that she and
her father should not attend the party hosted by the socially inferior
Cole family, but when an invitation does not arrive, she feels "that she
should like to have had the power of refusal" (E, p. 224). Discussing
Mr. Rushworth's plan to relandscape his estate, the thoughtful Edmund
Bertram says that he would want to make his own choices and not hire
a specialist improver: "[H]ad I a place to new fashion, I should not put
myself into the hands of an improver. I would rather have an inferior
degree of beauty, of my own choice, and acquired progressively. I would
rather abide by my own blunders than by his" (MP, pp. 66–67). In
contrast, the shallow Mary Crawford would "be most thankful to any
Mr. Repton who would undertake it, and give me as much beauty as he
could for my money; and I should never look at it, till it was complete."
Fanny, not yet decisive, is content to just watch: "It would be delightful
to *me* to see the progress of it all" (MP, p. 67).

The one time that being able to make a choice seems to be a bad thing
is when Fanny must choose what to wear to the ball. Fanny treasures
her amber cross ornament, a gift from her brother William, but needs
something to wear it with. Fanny must choose either the gold chain

given by Edmund or the gold necklace given by Mary Crawford. Fanny far prefers Edmund's chain, but Edmund asks Fanny to wear Mary's necklace because of his own interest in Mary. But "upon trial the one given her by Miss Crawford would by no means go through the ring of the cross. She had, to oblige Edmund, resolved to wear it—but it was too large for the purpose. His therefore must be worn; and having, with delightful feelings, joined the chain and the cross, those memorials of the two most beloved of her heart . . . she was able, without an effort, to resolve on wearing Miss Crawford's necklace too" (MP, pp. 314–15). Since the necklace from Mary is too large to go through the ring of the cross, Fanny has no choice and therefore can blamelessly wear Edmund's chain, which is what she wants. But once that point is settled, she then exercises her power of choice by deciding to wear Mary's necklace as well. Even when it seems better not to have to make a choice, Austen shows that another choice can make things better still.

Correspondingly, for Austen, people who cannot make choices deserve ridicule or worse. On a shopping trip, Marianne Dashwood can barely tolerate "the tediousness of Mrs. Palmer, whose eye was caught by every thing pretty, expensive, or new; who was wild to buy all, could determine on none, and dawdled away her time in rapture and indecision" (SS, pp. 187–88). Shopping with Emma, "Harriet, tempted by every thing and swayed by half a word, was always very long at a purchase" (E, p. 252); when Harriet cannot decide where her purchases should be delivered to, Emma impatiently directs, "That you do not give another half-second to the subject" and makes Harriet's decision for her (E, p. 254). More seriously, Mr. Weston's first wife, Miss Churchill, marries him in spite of her family's wishes, but she cannot completely give up the luxury of her parents' home at Enscombe, and thus "[t]hey lived beyond their income . . . [S]he did not cease to love her husband, but she wanted at once to be the wife of Captain Weston, and Miss Churchill of Enscombe" (E, p. 14). Miss Churchill cannot choose and dies in three years. For that matter, the ruinous affair between Maria Bertram and Henry Crawford results from their inability to make choices. Maria is unable to choose between married life and being pursued by Henry, and Henry is unable to choose between pursuing Fanny and overcoming Maria's coldness: "he could not bear to be thrown off by the woman whose smiles had been so wholly at his command. . . . [H]e went off with her at last, because he could not help it, regretting Fanny, even at the moment" (MP, pp. 541–42).

For Austen, choices bind. You can't have it both ways. Once you make a choice, you cannot pretend you did not make it. After visiting her friend Charlotte Lucas at her new married home with the vacant Mr. Collins, Elizabeth Bennet laments, "Poor Charlotte!—it was melancholy to leave her to such society!—But she had chosen it with her eyes open" (PP, p. 239). When Willoughby visits Marianne to seek forgiveness, even after having married Miss Grey for money, Elinor observes, "You are very wrong, Mr. Willoughby, very blamable... you ought not to speak in this way, either of Mrs. Willoughby or my sister. You had made your own choice. It was not forced on you" (SS, p. 373).

An inability to make choices can stem from a lack of resolution, which Austen consistently decries. When Emma and Mr. Knightley discuss why it has taken months for Frank Churchill to visit his father's new wife, Emma thinks that "his uncle and aunt will not spare him" and that Mr. Knightley does not understand "the difficulties of dependence" (E, pp. 156, 157). But Mr. Knightley firmly replies, "There is one thing, Emma, which a man can always do, if he chuses, and that is, his duty; not by manoeuvring and finessing, but by vigour and resolution" (E, p. 157). When Mr. Darcy chides his friend Mr. Bingley for yielding too quickly to the suggestions of others, Elizabeth at first defends Mr. Bingley: "You appear to me, Mr. Darcy, to allow nothing for the influence of friendship and affection" (PP, p. 54). However, when Mr. Bingley suspends without explanation his courtship of her sister Jane, Elizabeth correctly suspects Mr. Darcy's and Caroline Bingley's interference and reverses her position: "she could not think without anger, hardly without contempt, on that easiness of temper, that want of proper resolution, which now made him the slave of his designing friends" (PP, p. 151).

Austen hates encumbered choices. When Mrs. Norris asks Tom Bertram to help make a rubber for playing cards along with Dr. Grant and Mrs. Rushworth, Tom escapes by leading Fanny to dance. Tom complains to Fanny, "And to ask me in such a way too! without ceremony, before them all, so as to leave me no possibility of refusing! *That* is what I dislike most particularly. It raises my spleen more than any thing, to have the pretence of being asked" (MP, p. 141). Tom obliviously makes his point to Fanny, whose entire life is an encumbered choice. As Edwards (1965, p. 56) observes, "Robbing people of their choice lies at the heart of virtually every significant incident" in *Mansfield Park*. After it is discovered that Jane Fairfax regularly fetches her family's letters from the post office in the early morning even in the rain (to keep others from knowing about letters sent to her by Frank Churchill), Mrs. Elton tells Jane that she will ask the servant who fetches her own letters to fetch Jane's as well. Mrs. Elton odiously gives Jane no choice in the matter: "My dear Jane, say no more about it. The thing is determined.... [C]onsider that point as settled" (E, pp. 319–20). Finally, the entire plot of *Persuasion* is how Anne Elliot overcomes her original encumbered choice of refusing Captain Wentworth. Once Anne and Captain Wentworth finally understand their feelings for each other, they are "more exquisitely happy, perhaps, in their re-union, than when it had been first projected; more tender, more tried, more fixed in a knowledge of each other's character, truth, and attachment; more equal to act, more justified in acting" (P, p. 261). Now that Anne is older, no longer encumbered, "more equal to act," her choosing to marry Captain Wentworth does not merely correct her eight-year-old mistake; being empowered and unencumbered improves both choice and result.

Austen explains that to make a choice thoughtfully, you must understand the counterfactual of what would have happened had you chosen otherwise (in economics, this is the concept of "opportunity cost"). Mary Crawford argues that Edmund's chosen profession, the clergy, encourages indolence, not ambition. Edmund and Fanny disagree, but Mary cites as evidence her own brother-in-law Dr. Grant: "[T]hough he...often preaches good sermons, and is very respectable, *I* see him to be an indolent, selfish bon vivant, who must have his palate consulted in every thing, who will not stir a finger for the convenience of any one" (MP, p. 130). Edmund concedes ("Fanny, it goes against us. We cannot attempt to defend Dr. Grant"), but Fanny argues that in another profession Dr. Grant would have been even worse: "[W]hatever there may be to wish otherwise in Dr. Grant, would have been in a greater danger of becoming worse in a more active and worldly profession.... I have no doubt that he oftener endeavours to restrain himself than he would if he had been any thing but a clergyman" (MP, p. 131). Mary Crawford and Edmund might seem more strategically experienced than Fanny, but only Fanny understands that the relevant counterfactual is Dr. Grant having some other profession, not having no profession at all. When Edmund tells Fanny that he is concerned about Mary Crawford's character and is not sure whether to propose, perhaps Fanny would like to suggest that the relevant choice is not between Mary and not marrying at all, but between Mary and another young woman, such as herself. Similarly, when Emma wonders why the sophisticated Jane Fairfax spends so much time with the odious Mrs. Elton, Mrs. Weston offers, "We cannot suppose that she has any great enjoyment at the Vicarage, my dear Emma—but it is better than being always at home. Her aunt is a good creature, but, as a constant companion, must be very tiresome. We must consider what Miss Fairfax quits, before we condemn her taste for what she goes to" (E, p. 308).

Understanding the proper counterfactual, imagining all aspects of what would have happened had you chosen differently, is not always easy. When Anne meets Captain Wentworth's friends Captains Harville and Benwick, whose friendship is "so unlike the usual style of give-and-take invitations, and dinners of formality and display," Anne is forced to observe her counterfactual life had she not refused Captain Wentworth's original proposal: "'These would have been all my friends,' was her thought; and she had to struggle against a great tendency to lowness" (P, p. 105). Anne does know that some counterfactuals are not worth thinking about, namely those that have already been made impossible by events. During the concert in Bath, Mr. Elliot persists in his attentions toward Anne, making Captain Wentworth leave in a fit of jealousy. Anne feels goodwill toward Mr. Elliot, enough to raise the question of how Anne might think of him had there been no Captain Wentworth. But Anne knows that this counterfactual is irrelevant: "How she might have felt, had there been no Captain Wentworth in the case, was not worth enquiry; for there was a Captain Wentworth: and be the conclusion of the present suspense good or bad, her affection would be his for ever" (P, p. 208).

PREFERENCES

Game theory's assumption of numerical payoffs is essentially an assumption of commensurability, that complex mixtures of feelings can be reduced to a single sentiment. Austen acknowledges that this assumption can be problematic: for example, when Captain Wentworth arranges for Admiral and Mrs. Croft to take Anne Elliot home in their carriage, "it was a proof of his own warm and amiable heart, which she could not contemplate without emotions so compounded of pleasure and pain, that she knew not which prevailed" (P, p. 98). Regan (1997, p. 134) asks, "how can we compare, say a friendship and research on beetles? Such different things may both be valuable, but how can we possibly say one is more valuable than the other?"

But Austen consistently argues for commensurability. Austen almost always allows a mixture of feelings to resolve into a single feeling, usually just through the passage of time. At Fanny's first ball, Henry Crawford quickly engages Fanny for the first two dances, and "[h]er happiness on this occasion was very much à la mortal, finely chequered. To be secure of a partner at first, was a most essential good . . . but at the same time there was a pointedness in his manner of asking her, which she did not like, and she saw his eye glancing for a moment at her necklace— with a smile—she thought there was a smile—which made her blush and feel wretched" (MP, pp. 318–19). Fanny experiences both appreciation and revulsion, and "had no composure till he turned away to some one else. Then she could gradually rise up to the genuine satisfaction of having a partner, a voluntary partner" (MP, p. 319). Once Fanny is able to compose herself, she can reduce the two competing feelings into a single genuine satisfaction. After Henry Crawford tells Fanny that he secured her brother William's promotion and then proposes, Fanny is "in the utmost confusion of contrary feeling. She was feeling, thinking, trembling... agitated, happy, miserable, infinitely obliged, absolutely angry" (MP, pp. 349–50). But at the end of the day, the confusion is resolved because the pleasure lasts while the pain decays: "Fanny thought she had never known a day of greater agitation, both of pain and pleasure; but happily the pleasure was not of a sort to die with the day— for every day would restore the knowledge of William's advancement, whereas the pain she hoped would return no more" (MP, p. 356). When Edward Ferrars receives Lucy Steele's letter saying that she has married his brother, thereby releasing him from their engagement, he is "half stupified between the wonder, the horror, and the joy of such a deliverance" (SS, p. 413). Edward tells Elinor that "this is the only letter I ever received from her, of which the substance made me any amends for the defect of the style" (SS, p. 414). Even though Edward does not reduce his wonder, horror, and joy into a single satisfaction, he finds that disparate aspects, style and substance, can compensate for each other.

Austen actually delights in how one feeling can be compensated for by another feeling of a completely different kind. After Marianne is dumped by Willoughby, Mrs. Jennings tries to cheer her up, and Elinor "could have been entertained by Mrs. Jennings's endeavours to cure a disappointment in love, by a variety of sweetmeats and olives, and a good fire" (SS, p. 220). One might think that the trivial pleasures of food and physical warmth could not possibly compensate for a broken heart, but Mrs. Jennings has a point. Catherine Morland is disappointed that the Tilneys do not show up for their planned walk, even after the rain stops, and thus decides to go on a carriage ride with John Thorpe, Isabella, and her brother. "Catherine's feelings... were in a very unsettled state; divided between regret for the loss of one great pleasure, and the hope of soon enjoying another, almost its equal in degree, however unlike in kind.... To feel herself slighted by them was very painful. On the other hand, the delight of exploring an edifice like Udolpho, as her fancy represented Blaize Castle to be, was such a counterpoise of good as might console her for almost any thing" (NA, p. 85). Catherine experiences a mixture of two feelings quite unlike in kind, but one compensates more than fully for the other. Lady Russell dislikes visiting Sir Walter Elliot at his rental house in Bath because she dislikes Mrs. Clay, the guest of Sir Walter and his daughter Elizabeth. But Mr. Elliot, Sir Walter's presumptive heir, also often visits; Lady Russell begins to like him and think him a match for Anne, and "[h]er satisfaction in Mr. Elliot outweighed all the plague of Mrs. Clay" (P, p. 159). Sitting in her former schoolroom, Fanny recalls "the pains of tyranny, of ridicule, and neglect" but also the consoling kindnesses of Edmund, her aunt Lady Bertram, and her teacher Miss Lee, and "the whole was now so blended together, so harmonized by distance, that every former affliction had its charm" (MP, p. 178). Elizabeth Bennet relies upon commensurability: when Lady Catherine threatens that Mr. Darcy's family will despise her if she marries him, Elizabeth responds, "the wife of Mr. Darcy must have such extraordinary sources of happiness necessarily attached to her situation, that she could, upon the whole, have no cause to repine" (PP, p. 394).

A few times Austen uses explicitly quantitative analogies for happiness or sadness. When Henry Tilney leaves Northanger Abbey a few days early to prepare to receive Catherine and his sister and father at his own house at Woodston, he equates happiness with money that circulates in financial markets, remarking that "our pleasures in this world are always to be paid for, and that we often purchase them at a great disadvantage, giving ready-monied actual happiness for a draft on the future, that may not be honoured.... Because I am to hope for the satisfaction of seeing you at Woodston on Wednesday, which bad weather, or twenty other causes, may prevent, I must go away directly, two days before I intended it" (NA, p. 217). After Lydia Bennet runs off with Wickham unmarried, Elizabeth Bennet is certain that Mr. Darcy will therefore have nothing to do with her family; thus "had she known nothing of Darcy, she could have borne the dread of Lydia's infamy somewhat better. It would have spared her, she thought, one sleepless night out of two" (PP, p. 329). Dread can be measured in number of sleepless nights.

Given commensurability, a feeling's strength can be measured by how much of another feeling it takes to compensate for it. In economics, this is the concept of "compensating differential" (Rosen 1986); for example, Cornell graduating seniors require on average $13,037 more in yearly salary to work for Exxon as opposed to the Peace Corps (Frank 2004, p. 88). Mr. Darcy has underwritten Lydia's marriage and thus saved the Bennet family's reputation, but only Elizabeth is aware of her family's debt to him. When Mr. Bingley and Mr. Darcy unexpectedly visit, Mrs. Bennet snubs Mr. Darcy and showers attention on Mr. Bingley, even though Mr. Bingley has disregarded the family for months. "Elizabeth's misery increased, at such unnecessary, such officious attention!... At that instant she felt, that years of happiness could not make Jane or herself amends, for moments of such painful confusion." Elizabeth wishes never to see the two men again, saying to herself, "Their society can afford no pleasure, that will atone for such wretchedness as this!" But when Elizabeth sees Mr. Bingley's attraction to Jane, "the misery, for which years of happiness were to offer no compensation, received soon afterwards material relief, from observing how much the beauty of her sister re-kindled the admiration of her former lover" (PP, p. 373). Austen might be making fun of how the emotional strain of this moment is enough to make even Elizabeth's feelings fluctuate from one extreme to the other, but another interpretation is that Austen acknowledges the possibility of incommensurability only to firmly rule in favor of commensurability. A lover's admiration, if it leads to marriage, could very well result in years of happiness to a bride's sister, even under the soberest calculation.

Similarly, when Edmund Bertram and Mary Crawford leave Fanny by herself in the Sotherton grounds, her "best consolation was in being assured that Edmund had wished for her very much ... but this was not quite sufficient to do away with the pain of having been left a whole hour, when he had talked of only a few minutes" (MP, p. 120). Finally, after Maria Bertram's affair and subsequent divorce, Maria goes off to live with Mrs. Norris in a remote location, and "Mrs. Norris's removal from Mansfield was the great supplementary comfort of Sir Thomas's life. ... To be relieved from her ... was so great a felicity that ... there might have been danger of his learning almost to approve the evil which produced such a good" (MP, pp. 538–39). For Sir Thomas, the pleasure of a life without Mrs. Norris is large enough to nearly counterpoise the collapse of his family's reputation.

REVEALED PREFERENCES

The strength of competing feelings is best revealed by a person's choice; this is the idea of "revealed preference" in economics (see for example Varian 2006). After Jane Bennet receives a letter from Mr. Bingley's sister Caroline saying that her brother will most likely marry Mr. Darcy's sister Georgiana, Jane asks Elizabeth, "[C]an I be happy, even supposing the best, in accepting a man whose sisters and friends are all wishing him to marry elsewhere?" Elizabeth replies, "You must decide for yourself ... and if upon mature deliberation, you find that the misery of disobliging his two sisters is more than equivalent to the happiness of being his wife, I advise you by all means to refuse him" (PP, p. 134). Similarly, Elizabeth can estimate the strength of Mr. Darcy's feelings given his decision to propose: "so much in love as to wish to marry her in spite of all the objections which had made him prevent his friend's marrying her sister ... was almost incredible! it was gratifying to have inspired unconsciously so strong an affection" (PP, p. 216). Even if an affection is inspired unconsciously, a person's decision reveals its strength.

If there is any doubt about a person's preferences, his choices provide proof. Frank Churchill tells Emma that Jane Fairfax must be an excellent pianist because Mr. Dixon, even while in love with another woman, "would yet never ask that other woman to sit down to the instrument, if the lady in question [Jane] could sit down instead....That, I thought, in a man of known musical talent, was some proof" (E, p. 217). When Marianne agrees to go along with Mrs. Jennings to London, on the chance that she might see Willoughby there, Elinor observes, "That Marianne, fastidious as she was, thoroughly acquainted with Mrs. Jennings' manners, and invariably disgusted by them...should disregard whatever must be most wounding to her irritable feelings, in her pursuit of one object, was such a proof, so strong, so full of the importance of that object to her" (SS, p. 176).

Preferences can be revealed even by a hypothetical choice. After Catherine realizes that John Thorpe manipulated her into abandoning her plan to walk with Eleanor and Henry Tilney and instead ride with him to Blaize Castle, "rather than be thought ill of by the Tilneys, she would willingly have given up all the happiness which its walls could supply" (NA, p. 86). Even after his brother Robert is bestowed the estate that Edward Ferrars would have received, happy Edward is "free from every wish of an exchange" (SS, p. 428). Anne admires the Musgrove sisters' happiness, but "would not have given up her own more elegant and cultivated mind for all their enjoyments" (P, p. 43). When Edmund Bertram bids a final goodbye to Mary Crawford, the depth of Fanny's satisfaction is such that "there are few who might not have been glad to exchange their greatest gaiety for it" (MP, p. 533). When Emma wonders if she is in love with Frank Churchill, she first examines her feelings: "This sensation of listlessness, weariness, stupidity, this disinclination to sit down and employ myself, this feeling of every thing's being dull and insipid about the house!—I must be in love; I should be the oddest creature in the world if I were not—for a few weeks at least" (E, p. 283). But what is quite conclusive is what she would choose in different scenarios: "forming a thousand amusing schemes for the progress and close of their attachment, fancying interesting dialogues, and inventing elegant letters; the conclusion of every imaginary declaration on his side was that she *refused him*" (E, p. 284).

A common objection to the payoff maximization model is that two alternatives can be different in so many ways that they cannot be directly compared. Anderson (1997, p. 99) asks, "Is Henry Moore's sculpture, *Recumbent Figure*, as intrinsically good a work of art as Chinua Achebe's novel, *Things Fall Apart*? What could be the point of answering such a question?" But Austen likes direct comparisons. Mary Crawford had liked Tom Bertram, but once he returns to Mansfield after an absence, listening to him talk is enough "to give her the fullest conviction, by the power of actual comparison, of her preferring his younger brother" (MP, p. 134). After Emma is engaged with Mr. Knightley, she converses with Frank Churchill, and "falling naturally into a comparison of the two men, she felt, that pleased as she had been to see Frank Churchill...she had never been more sensible of Mr. Knightley's high superiority of character" (E, p. 524). Edward Ferrars foolishly engaged himself to Lucy Steele because "Lucy appeared everything that was amiable and obliging.... I had seen so little of other women, that I could make no comparisons, and see no defects" (SS, p. 411). Sometimes only through comparisons can an alternative be truly appreciated.

NAMES FOR STRATEGIC THINKING

Austen uses specific terms to refer to strategic thinking, including "penetration," "foresight," and "sagacity." For example, Mr. Bennet tells Elizabeth, "Young ladies have great penetration in such matters as these; but I think I may defy even *your* sagacity, to discover the name of your admirer" (PP, p. 401). When Marianne wants to stay home instead of joining the family to visit Lady Middleton, Mrs. Dashwood concludes that she must be planning to receive a visit from Willoughby. "On their return from the park they found Willoughby's curricle and servant in waiting at the cottage, and Mrs. Dashwood was convinced that her conjecture had been just. So far it was all as she had foreseen" (SS, p. 87). Along with "foresight," the term "penetration" analogizes strategic thinking as vision. Isabella Thorpe tells Catherine that she must have already known that James would propose: "Yes, my dear Catherine, it is so indeed; your penetration has not deceived you.—Oh! that arch eye of yours!—It sees through every thing" (NA, p. 119). Similarly, an obsolete meaning of "sagacity" is a keen sense of smell.

After Mr. Elton's surprise proposal in the carriage, Emma recalls that "[t]o Mr. John Knightley was she indebted for her first idea on the subject, for the first start of its possibility. There was no denying that those brothers had penetration" (E, p. 146). Mr. John Knightley, "with some slyness," had warned a skeptical Emma, "[H]e seems to have a great deal of good-will towards *you*.... You had better look about you, and ascertain what you do, and what you mean to do" (E, p. 120). Mr. John Knightley rides in the same carriage together with Emma and Mr. Elton on the way to Randalls, but on the ride back home, "forgetting that he did not belong to their party, stept in after his wife" in her carriage, leaving the two of them alone together (E, p. 139). Perhaps Mr. John Knightley "forgets" purposefully, anticipating that Mr. Elton's actions would validate his warning to Emma. A person with penetration discerns another's preferences and can also strategically make those preferences reveal themselves.

There are more than fifty strategic plans specifically named as "schemes" in Austen's six novels; Edwards (1965, p. 55) writes that "meddling" is the theme of Austen's novels and "indeed, it is the theme of most classic fiction and many of our difficulties with life." For example, after Elizabeth declines Mr. Collins's proposal, Elizabeth thanks Charlotte Lucas for talking with him, but "Charlotte's kindness extended farther than Elizabeth had any conception of;—its object was nothing less, than to secure her from any return of Mr. Collins's addresses, by engaging them towards herself. Such was Miss Lucas's scheme" (PP, p. 136). Occasionally "scheme" is used less specifically to mean a social engagement; for example, when Lydia is in London with Wickham, she says, "I did not once put my foot out of doors, though I was there a fortnight. Not one party, or scheme, or any thing" (PP, p. 352). But of course social engagements often have strategic intentions; after Marianne and Willoughby first meet, "the schemes of amusement at home and abroad, which Sir John had been previously forming, were put into execution. The private balls at the park then began. . . . In every meeting of the kind Willoughby was included (SS, p. 63). After their engagement, Mr. Collins and Charlotte spend a week "in professions of love and schemes of felicity" (PP, p. 158), and indeed, a strategic plan is similar to a dream, a scenario in which everyone's actions fall exactly into place.

The term "contrive" is used similarly although less often. For example, when Emma and Harriet walk by Mr. Elton's vicarage and Harriet is curious to see it, Emma says, "I wish we could contrive it . . . but I cannot think of any tolerable pretence for going in" (E, p. 90). "Art" is another term for strategic manipulation, as in Marianne's exclamation when learning that Willoughby is engaged to another: "And yet this woman—who knows what her art may have been—how long it may have been premeditated, and how deeply contrived by her!" (SS, p. 216). "Art" has connotations of persuasion, as when Lady Catherine warns Elizabeth to stay away from Mr. Darcy: "*your* arts and allurements may, in a moment of infatuation, have made him forget what he owes to himself and to all his family. You may have drawn him in" (PP, pp. 392–93). "Sly" has associations with concealment, as when Mrs. Gardiner writes to Elizabeth that Mr. Darcy secretly paid off Wickham to marry Lydia: "I thought him very sly;—he hardly ever mentioned your name. But slyness seems the fashion" (PP, p. 360).

"Cunning" has admittedly negative connotations; for example, after Mrs. Smith tells Anne about how Mr. Elliot has wormed his way into the family to prevent the marriage of Anne's father Sir Walter with Mrs. Clay, Anne replies, "There is always something offensive in the details of cunning" (P, p. 224). None of the other terms, however, necessarily indicates disapproval. For example, in order to propose to Charlotte, Mr. Collins sneaks out of the Bennet residence "with admirable slyness" (PP, p. 136). Suspecting General Tilney of having murdered or imprisoned his wife, Catherine, "blushing at the consummate art of her own question," asks Eleanor where Mrs. Tilney's picture is hung (NA, p. 185).

Sometimes, the term "calculate" is used for strategic thinking; for example, to recommend Willoughby, Sir John Middleton's parties "were exactly calculated to give increasing intimacy to his acquaintance with the Dashwoods" (SS, p. 63). "Calculation" of course has mathematical connotations. A common objection to game theory is that people surely do not calculate as they are assumed to do in mathematical models; for example, Elster (2007, p. 5) writes, "Do real people act on the calculations that make up many pages of mathematical appendixes in leading journals? I do not think so."

But in Austen's novels, people calculate all the time without the slightest intimation that calculation is difficult, "cold," or unnatural. Willoughby offers Marianne "a horse, one that he had bred himself on his estate in Somersetshire, and which was exactly calculated to carry a woman" (SS, p. 68). His rival Colonel Brandon can calculate, too: when Marianne falls ill, he offers to retrieve Mrs. Dashwood, and "with all the firmness of a collected mind, made every necessary arrangement with the utmost dispatch, and calculated with exactness the time in which she might look for his return" (SS, p. 352). Colonel Brandon's calculation, calm but not unemotional, provides warm assurance. Similarly, since Mr. Collins is heir to Mr. Bennet's property, after he is engaged to her daughter Charlotte, Lady Lucas "began directly to calculate, with more interest than the matter had ever excited before, how many years longer Mr. Bennet was likely to live" (PP, p. 137). The rapidity of her calculation is an expression of her joy. After Louisa's fall, Captain Wentworth realizes to his surprise that others consider him engaged to her, and retreats to his brother's house "lamenting the blindness of his own pride, and the blunders of his own calculations" (P, p. 264). Captain Wentworth blunders, but not because calculating is difficult or unnatural. Captain Harville asks Anne to feel what a man feels when he is apart from his family, "when, coming back after a twelvemonth's absence perhaps, and obliged to put into another port, he calculates how soon it be possible to get them there, pretending to deceive himself, and saying, 'They cannot be here till such a day,' but all the while hoping for them twelve hours sooner" (P, p. 255). Calculation is as human as managing one's own expectations but at the same time hoping against them.

By the way, Austen uses the term "rational," but in the unspecialized sense of "reasonable" or "practical," as when Mrs. Croft tells her brother Captain Wentworth that women can be perfectly comfortable on a man-of-war: "I hate to hear you talking so, like a fine gentleman, and as if women were all fine ladies, instead of rational creatures" (P, p. 75). After Elizabeth refuses Mr. Collins's proposal, he tells her, "I shall chuse to attribute it to your wish of increasing my love by suspense" and Elizabeth replies, "Do not consider me now as an elegant female intending to plague you, but as a rational creature speaking the truth from her heart" (PP, p. 122). Here "rational" means unstrategic. Austen uses a variant of "strategic" only once. When Elizabeth hears that Mr. Darcy was somehow involved with Lydia's marriage, she asks her aunt Mrs. Gardiner to explain: "[M]y dear aunt, if you do not tell me in an honourable manner, I shall certainly be reduced to tricks and stratagems to find it out" (PP, p. 354).

More pretentious terms for strategic thinking, such as "manoeuvre," "taking in," "catching," and "setting one's cap," are used mainly by people who do not really know what they are doing. Mary Crawford says that marriage is "a manoeuvring business. I know so many who have married in the full expectation and confidence of some ... good quality in the person, who have found themselves entirely deceived. ... What is this, but a take in?" (MP, pp. 53–54). But Mary's sister Mrs. Grant corrects her: "[I]f one scheme of happiness fails, human nature turns to another; if the first calculation is wrong, we make a second better; we find comfort somewhere" (MP, p. 54). Note that if one calculation fails, a couple is saved not by hope and faith but by a second calculation. When Sir John says that Willoughby is "very well worth catching," Mrs. Dashwood replies, "I do not believe ... that Mr. Willoughby will be incommoded by the attempts of either of *my* daughters towards what you call *catching him*. It is not an employment to which they have been brought up" (SS, pp. 52–53). After Marianne asks how Willoughby dances, Sir John continues, "You will be setting your cap at him now." Marianne, now provoked, objects: "That is an expression, Sir John ... which I particularly dislike. I abhor every common-place phrase by which wit is intended; and 'setting one's cap at a man,' or 'making a conquest,' are the most odious of all." Sir John "did not much understand this reproof" and cannot shut up, continuing, "Ay, you will make conquests enough, I dare say, one way or other. Poor Brandon! he is quite smitten already, and he is very well worth setting your cap at" (SS, pp. 53–54). Sir John thinks he knows the lingo of strategic thinking, but displays his imbecility in his nonresponse to Marianne's direct objection. When Marianne asks Sir John what kind of man Willoughby is, Sir John cannot "describe to her the shades of his mind" (SS, p. 52).

STRATEGIC SOPHOMORES

Indeed, poor strategic thinking is best illustrated not by the naive but by the sophomoric, people who think they know something but don't. Like Sir John, the easily duped John Dashwood fancies himself knowledgeable about the strategic ways of women, instructing his sister Elinor to capture Colonel Brandon: "[S]ome of those little attentions and encouragements which ladies can so easily give, will fix him, in spite of himself" (SS, p. 254). Similarly, Mr. Collins thinks himself expert when Elizabeth Bennet refuses his proposal: "I am not now to learn...that it is usual with young ladies to reject the addresses of the man whom they secretly mean to accept, when he first applies for their favour" (PP, p. 120).

Mrs. Jennings is Austen's archetypical sophomore, who "missed no opportunity of projecting weddings among all the young people of her acquaintance" (SS, p. 43). After Willoughby clandestinely takes Marianne to see Allenham Court, which he hopes to inherit, Mrs. Jennings proudly reports, "I have found you out in spite of all your tricks. I know where you spent the morning....I hope you like your house, Miss Marianne" (SS, p. 79). In London, Willoughby does not respond to Marianne's letters, and acts with brutal coldness to her at a party. A letter arrives from him, and Elinor "felt immediately such a sickness at heart as made her hardly able to hold up her head, and sat in such a general tremour as made her fear it impossible to escape Mrs. Jennings's notice." But Mrs. Jennings is comfortably unaware: "Of Elinor's distress, she was too busily employed in measuring lengths of worsted for her rug, to see any thing at all" (SS, p. 206). Later Mrs. Jennings says she could not have possibly known that Willoughby had written to tell Marianne that he is marrying Miss Grey: "But then you know, how should I guess such a thing? I made sure of its being nothing but a common love letter, and you know young people like to be laughed at about them" (SS, pp. 221–22). When Colonel Brandon visits, Mrs. Jennings whispers to Elinor, "The Colonel looks as grave as ever you see. He knows nothing of it; do tell him, my dear," but she is wrong as usual; Colonel Brandon has already heard that Willoughby will marry Miss Grey (SS, p. 225). Later Mrs. Jennings stealthily observes Colonel Brandon talking quietly with Elinor: "she could not keep herself from seeing that Elinor changed colour, attended with agitation, and was too intent on what he said, to pursue her employment....[H]e seemed to be apologizing for the badness of his house. This set the matter beyond a doubt" (SS, p. 318). But instead of offering marriage, Colonel Brandon is merely talking to Elinor about offering a living to Edward Ferrars. After Elinor explains, Mrs. Jennings's misunderstanding provides "considerable amusement for the moment, without any material loss of happiness to either, for

Mrs. Jennings only exchanged one form of delight for another, and still without forfeiting her expectation of the first" (SS, p. 331). Even proved wrong, Mrs. Jennings persists in supposing that Elinor and Colonel Brandon will marry.

Sophomores' strategic plans backfire. Sir William Lucas, Charlotte Lucas's father, "struck with the action of doing a very gallant thing," tries to get Elizabeth and Mr. Darcy to dance together but fails (PP, p. 28). When they do dance together on a later occasion, Sir William thinks himself clever when he says to them, "I must hope to have this pleasure often repeated, especially when a certain desirable event, my dear Eliza, (glancing at her sister and Bingley,) shall take place" (PP, p. 104). Mr. Darcy reads Sir William's glance, and realizing now that his friend Mr. Bingley is in real danger of marrying Jane, persuades Mr. Bingley against it, which in turn makes Elizabeth despise him. Sir William's plan backfires, and indeed having a plan backfire demonstrates your strategic ineptitude better than having no plan at all. At Pemberley, in the presence of Mr. Darcy and his sister Georgiana, Caroline Bingley asks Elizabeth a question which alludes to Wickham, in order "to discompose Elizabeth, by bringing forward the idea of a man to whom she believed her partial, to make her betray a sensibility which might injure her in Darcy's opinion." But Caroline Bingley is unaware of Georgiana's and Mr. Darcy's painful association with Wickham, who had tried to elope with Georgiana when she was fifteen. Elizabeth, who knows this history, is able to respond calmly, and "the very circumstance which had been designed to turn his thoughts from Elizabeth, seemed to have fixed them on her more, and more cheerfully" (PP, p. 298).

Austen's strategic sophomores take pride in the trivial. When Mrs. Allen says that Henry Tilney "is a very agreeable young man," Mrs. Thorpe replies, "Indeed he is, Mrs. Allen... I must say it, though I *am* his mother, that there is not a more agreeable young man in the world." Mrs. Allen is smart enough to figure it out, "for after only a moment's consideration, she said, in a whisper to Catherine, 'I dare say she thought I was speaking of her son'" (NA, p. 54). Mrs. Norris suspects that Dick Jackson, who brings some boards to his father as the servants' meal is starting, has designs on a free meal, and takes pride in foiling him: "as I hate such encroaching people... I said to the boy directly—(a great lubberly fellow of ten years old you know, who ought to be ashamed of himself,) *I'll* take the boards to your father, Dick; so get you home again as fast as you can'" (MP, p. 167). Mrs. Norris, a moocher herself, is proud of outwitting a ten-year-old child. Similarly, one of Charlotte Lucas's younger brothers declares, "If I were as rich as Mr. Darcy... I would... drink a bottle of wine a day." Mrs. Bennet answers, "[I]f I were to see you at it, I should take away your bottle directly" and "[t]he boy protested that she should not; she continued to declare that she would, and the argument ended only with the visit" (PP, p. 22). Mrs. Bennet can only fight to a draw with a child. Before Jane Fairfax's engagement with Frank Churchill is made public, Mrs. Elton is very proud to be in on the secret; when Emma visits Jane, she sees Mrs. Elton "with a sort of anxious parade of mystery fold up a letter which she had apparently been reading aloud to Miss Fairfax.... 'I mentioned no *names*, you will observe.—Oh! no; cautious as a minister of state. I managed it extremely well.'... It was a palpable display, repeated on every possible occasion" (E, p. 495). Mrs. Elton is not more ridiculous than Emma herself, when, eager for validation and praise, she cannot help letting Frank Churchill in on her arch speculation that Mr. Dixon was Jane Fairfax's secret lover: "it had been so strong an idea, that it would escape her, and his submission to all that she told, was a compliment to her penetration" (E, p. 249). At least Emma upon marriage tries to graduate from the sophomoric, while Mrs. Elton remains.

EYES

"Penetration" and "foresight" involve vision, but eyes reveal as well as detect preferences, as discussed in chapter 2. Mrs. Jennings thinks that Colonel Brandon is interested in Elinor because of "his open pleasure in meeting her after an absence of only ten days, his readiness to converse with her, and his deference for her opinion" (SS, pp. 345–46). But Elinor "watched his eyes, while Mrs. Jennings thought only of his behaviour;—and while his looks of anxious solicitude on Marianne's feeling[s]... because unexpressed by words, entirely escaped the latter lady's observation;—*she* could discover in them the quick feelings, and needless alarm of a lover" (SS, p. 346). Eyes speak louder than behavior or words. At Fanny's first ball, Lady Bertram and Mrs. Norris ascribe her good looks to her dress, but Sir Thomas knows the true estimation of her beauty: "when they sat down to table the eyes of the two young men assured him" (MP, p. 316). Once Henry Crawford proposes, Edmund can see the reluctance in Fanny's eyes: "Fanny was worth it all... but he did not think he could have gone on himself with any woman breathing, without something more to warm his courage than his eyes could discern in hers" (MP, p. 388).

Mr. Knightley's suspicion that Frank Churchill admires Jane Fairfax and not Emma Woodhouse starts when "he had seen a look, more than a single look, at Miss Fairfax, which, from the admirer of Miss Woodhouse, seemed somewhat out of place" (E, p. 373). Mr. Knightley "suspected in Frank Churchill the determination of catching her eye" and later positions himself "to see as much as he could, with as little apparent observation" (E, pp. 376–77). Butte (2004, p. 120) notes that "[h]is ability to track people's perceptions and their perceptions of how others perceive them is indeed the core of Knightley's wisdom." Mr. Knightley, the look-reading specialist, has *nunchi* (explained in chapter 2). In contrast, Sir William Lucas compliments Elizabeth and Mr. Darcy on their dancing, saying, "You will not thank me for detaining you from the bewitching converse of that young lady, whose bright eyes are also upbraiding me," but Elizabeth has yet to like Mr. Darcy and had agreed to dance quite reluctantly (PP, p. 104). Sir William Lucas does not have *nunchi.*

Looking at someone's eyes is powerfully informative, but Austen admits its limitations. The day after Anne talks with Captain Wentworth at the concert in Bath, her friend Mrs. Smith tells her, "You need not tell me that you had a pleasant evening. I see it in your eye. . . . Your countenance perfectly informs me that you were in company last night with the person, whom you think the most agreeable in the world." Anne blushes "in the astonishment and confusion excited by her friend's penetration, unable to imagine how any report of Captain Wentworth could have reached her" (P, p. 210). Mrs. Smith mistakenly thinks that Mr. Elliot is the most agreeable person; seeing Anne's eye, Mrs. Smith is completely correct about Anne's feelings, but wrong about their object. When Colonel Brandon visits Marianne while she is ill, Elinor "soon discovered in his melancholy eye and varying complexion as he looked at her sister, the probable recurrence of many past scenes of misery to his mind, brought back by that resemblance between Marianne and Eliza already acknowledged, and now strengthened by the hollow eye, the sickly skin, the posture of reclining weakness" (SS, p. 385). Eliza was an orphan and Colonel Brandon's first love when they were both children. Her guardian, Colonel Brandon's father, forced her to marry Colonel Brandon's brother, who did not love her, and after her divorce she fell into poverty and illness; Colonel Brandon had taken care of her in her last days of life. Colonel Brandon had told Elinor this story in order to better explain Willoughby's character: Willoughby had seduced Eliza's daughter, also named Eliza, and fathered a bastard child with her. Elinor reads Colonel Brandon's melancholy eye, but her deeper understanding of his mind is based on knowledge of his past experiences. In contrast, "Mrs. Dashwood, not less watchful of what passed than her daughter, but with a mind very differently influenced, and therefore watching to very different effect, saw nothing in the Colonel's behaviour but what arose from the most simple and self-evident sensations" (SS, pp. 385–86). Mrs. Dashwood can read eyes as well as Elinor can, but does not know the back story.

CHAPTER SEVEN

Austen's Competing Models

AUSTEN'S EMPHASIS on choice and strategic thinking does not keep her from considering competing models of human behavior; she acknowledges their relevance while maintaining a game-theoretic worldview overall. Indeed, strategic thinking is best understood in contrast and interaction with competing models.

EMOTIONS

One competing model focuses on people's emotions. Austen acknowledges that emotions can cause bad decisions. For example, the jealous Caroline Bingley remarks to Mr. Darcy that Elizabeth's eyes have "a sharp, shrewish look," but this only makes him reply that Elizabeth is "one of the handsomest women of my acquaintance" (PP, pp. 299, 300). Thus "Miss Bingley was left to all the satisfaction of having forced him to say what gave no one any pain but herself" (PP, p. 300). Influenced by emotion, Miss Bingley does not think her plan through and it backfires: "angry people are not always wise" (PP, p. 299). As Loewenstein (2000, p. 428) writes, "immediate visceral factors can have an enormous influence on behavior in the absence of cognitive deliberations.... When behavior is driven by intense visceral factors, it stretches the meaning of the term to say that people are making 'decisions.' "

But Austen's heroines, when overwhelmed by emotions, still make good choices. They choose well in terms of both result and process, using "good sense." Their emotions do not hinder as much as drive their decisions; emotion can sharpen decision-making as well as cloud it. After Edmund Bertram gives Fanny the gold chain as "a token of the love of one of your oldest friends," he starts to walk away (MP, p. 303). Fanny, not wanting the moment to end, "overpowered by a thousand feelings of pain and pleasure, could attempt to speak; but quickened by one sovereign wish she then called out, 'Oh! cousin, stop a moment, pray stop' " (MP, pp. 303–34). Her feelings overpower her, but her wish is sovereign. Similarly, when Henry Crawford unexpectedly shows up at Fanny's house at Portsmouth, "[g]ood sense, like hers, will always act when really called upon" (MP, p. 463). A petrified Fanny manages to introduce him to her mother as a friend of her brother William, not as someone who is seeking her hand, even though "the terrors that occurred of what this visit might lead to were overpowering, and she fancied herself on the point of fainting away" (MP, p. 464). When Mr. Knightley confesses his love to Emma, "the expression of his eyes overpowered her" and she is "almost ready to sink under the agitation of this moment" (E, pp. 468–69). But "[w]hile he spoke, Emma's mind was most busy, and, with all the wonderful velocity of thought, had been able—and yet without losing a word—to catch and comprehend the exact truth of the whole; to see that...Harriet was nothing; that she was every thing herself" (E, p. 469). Emma's feelings do not hinder but increase the velocity of her cognition.

When Lucy Steele tells Elinor that she is secretly engaged to Edward Ferrars, Elinor's "security sunk; but her self-command did not sink with it" (SS, p. 151). Later, when Elinor reveals this to Marianne, who has been moaning over Willoughby, Marianne emotes, "How barbarous have I been to you!—you, who have been my only comfort, who have borne with me in all my misery, who have seemed to be only suffering for me!" (SS, p. 299). "[T]enderest caresses followed this confession," and the emotion of this moment is no less strong for being tender instead of overpowering (SS, p. 300). But Elinor knows that this is exactly the right moment to make her sister promise to behave: "In such a frame of mind as she was now in, Elinor had no difficulty in obtaining from her whatever promise she required; and at her request, Marianne engaged never to speak of the affair to any one with the least appearance of bitterness" (SS, p. 300). When Marianne "with feverish wildness" calls out for her mother, Elinor's "alarm increased so rapidly, as to determine her on sending instantly for Mr. Harris, and dispatching a messenger to Barton for her mother" (SS, p. 351). Marianne had been wanting to go home for months, and had been sick for several days with Elinor passively trusting in rest and the apothecary. Elinor's feeling of alarm is what finally kicks her into decisive action. Mrs. Dashwood is usually as emotionally demonstrative as Marianne, and when she arrives Elinor worries that she will make the most of the opportunity and disturb Marianne's sleep, but even Mrs. Dashwood "could be calm, could be even prudent, when the life of a child was at stake" (SS, p. 378).

Captain Wentworth tells Anne that he cannot understand how Captain Benwick could transfer his affection so easily to Louisa from his late fiancée Fanny Harville: "A man does not recover from such a devotion of the heart to such a woman!—He ought not—he does not." Anne "was struck, gratified, confused, and beginning to breathe very quick, and feel an hundred things in a moment. It was impossible for her to enter on such a subject." His warm expression of the constancy of a man's devotion cuts too close; Anne cannot address it directly, but does not want to change the subject completely. Burdened by a hundred feelings, she is still able to think quickly and thus "only deviated so far as to say—'You were a good while at Lyme, I think?' " (P, p. 199). Of course, as discussed in chapter 5, Anne's ability to think strategically under intense emotional strain is demonstrated by her skillful emergency dispatch at Lyme and her contingency planning, even while overpowered by Captain Wentworth's letter, to make sure that she will see him again. After General Tilney kicks out Catherine Morland from Northanger Abbey for not being an heiress, Catherine and Elinor Tilney share "a long and affectionate embrace [which] supplied the place of language in bidding each other adieu," but the intense sadness and awkwardness of this moment do not keep Catherine from trying to maintain contact with her object, Henry Tilney: "with quivering lips [she] just made it intelligible that she left 'her kind remembrance for her absent friend' " (NA, p. 237).

What about Austen's men? Mr. Darcy recalls that after Elizabeth had refused his proposal, "I was angry perhaps at first, but my anger soon began to take a proper direction" (PP, p. 410). His anger does not interfere with but rather becomes purposeful action. According to Solomon (2003, pp. 146–47), "emotions are purposive.... [T]hey are *in themselves* strategic and political.... [E]motions do not just 'happen' to us, as the whole language of 'passion' and 'being struck by' would suggest. They are, with some contentious stretching of the term, activities that we 'do,' stratagems that work for us, both individually and collectively." After Edmund says goodbye to Mary Crawford for the last time and walks away, Mary calls out to him with a "saucy playful smile, seeming to invite, in order to subdue me," but as Edmund later tells Fanny, "I resisted; it was the impulse of the moment to resist, and still walked on. I have since—sometimes—for a moment—regretted that I did not go back; but I know I was right" (MP, p. 531). An impulsive feeling can help you make the right choice. According to Pessoa (2008, p. 148), "emotion and cognition not only strongly interact in the brain... they are often integrated so that they jointly contribute to behaviour."

Just as intense emotion does not necessarily lead to poor choices, calmness does not necessarily lead to good ones. For example, Elinor observes that Lady Middleton's "reserve was a mere calmness of manner with which sense had nothing to do" (SS, p. 65). Austen's characters who lack true feeling tend to be bad, not good, at strategic thinking. When Maria Bertram weds, "her mother stood with salts in her hand, expecting to be agitated—her aunt [Mrs. Norris] tried to cry" (MP, p. 237). When Mrs. Norris's husband dies, she "consoled herself for the loss of her husband by considering that she could do very well without him" (MP, p. 26). Lady Bertram, through inexertion, and Mrs. Norris, through a more active ignorance, lack both feelings and strategic sense.

Emotions can affect people's choices, but people can also strategically manage their emotions. The simplest way, Austen notes, is to take some time out. After Edward Ferrars tells the Dashwoods that Lucy Steele has married his brother and not him, he sees Elinor run out of the room in tears, and thus falls into a reverie; "without saying a word, [he] quitted the room, and walked out towards the village——leaving the others in the greatest astonishment" (SS, p. 408). He takes a few hours, and having "walked himself into the proper resolution," is able to return and propose to Elinor (SS, p. 409). Similarly, when Henry Tilney confronts his father about his mistreatment of Catherine, "[t]he general was furious in his anger, and they parted in dreadful disagreement. Henry, in an agitation of mind which many solitary hours were required to compose," decides to travel the next day to the Morland home to apologize and propose (NA, p. 257). When Edmund tells Fanny in her room upstairs that she is one of "the two dearest objects I have on earth," Fanny tries "to tranquillise herself as she could," by repeating to herself that even if Mary Crawford is first in his affections, she herself is second (MP, p. 306). She tells herself that she wishes only that Edmund could see Mary's faults as clearly as she can; Fanny does not dare think she might deserve him herself. But all this reasoning is not enough, so in addition Fanny moons over the handwritten note Edmund has just left for her. Finally, "[h]aving regulated her thoughts and comforted her feelings by this happy mixture of reason and weakness, she was able, in due time, to go down and resume her usual employments" (MP, p. 308). According to Zelazo and Cunningham (2007, p, 136), "*Emotion regulation* can occur in a variety of ways ... but one of the most obvious varieties is the deliberate self-regulation of emotion via conscious cognitive processing."

Surely some emotional responses are not subject to conscious control. Mullan (2012, p. 259) notices that "one form of expression that is missing from all those dramatisations of Austen's novels ... is the blush. Weeping is easy for any accomplished performer, but the Austen blush— that most truly involuntary signal of feeling—is almost impossible." Blushing, commonly understood as an autonomic response to feelings such as embarrassment, is the prototypical example of an action directly caused by emotion: "As an involuntary reflex seen by others, blushing 'leaks' an emotional condition" (Shearn, Bergman, Hill, Abel, and Hinds 1992, p. 431). For example, when Elizabeth and Mr. Darcy meet unexpectedly at Pemberley, "[t]heir eyes instantly met, and the cheeks of both were overspread with the deepest blush" (PP, p. 278). Because it is presumably unwilled, blushing is supposed to truthfully reveal a person's feelings (Frank 1988). For example, when Henry Crawford reads Shakespeare aloud, Fanny tries hard to ignore him but cannot, and her reaction, "blushing and working as hard as ever" on her needlework, "had been enough to give Edmund encouragement for his friend" (MP, p. 390).

But even blushing is sometimes understood by Austen as partly a matter of choice. For example, after James Morland is dumped by his fiancée Isabella Thorpe, he writes to his sister Catherine, closing with the line, "Dearest Catherine, beware how you give your heart" (NA, p. 208). Reading the letter, Catherine is distraught and a concerned Henry Tilney wants to know why; Catherine almost hands the letter to him so he can read it himself, but hesitates, "recollecting with a blush the last line." Henry suggests that Catherine might read aloud only the sections that are not private, but Catherine reconsiders: " 'No, read it yourself,' cried Catherine, whose second thoughts were clearer. 'I do not know what I was thinking of,' (blushing again that she had blushed before,)— 'James only means to give me good advice' " (NA, p. 210). Catherine first blushes because she thinks that Henry might think she intends to suggest something by handing him a letter that mentions giving her heart; by blushing, her innocence is made evident. But thinking again, Catherine is embarrassed that the need to blush had even occurred to her; an entirely innocent young woman would never have thought it possible to use a brother's advice to flirt. Using Halsey's (2006, p. 232) terminology, Catherine thought that her first blush was a "transparently honest" blush but blushes again because she realizes that Henry may have understood it as a "slyly knowing" blush. Instead of blushing for blushing, Catherine would rather not have blushed in the first place. Catherine's first blush is explained by her emotion of embarrassment. Her second blush might be similarly explained, even though a cognitive element must be included (her figuring out that her first blush is itself cause for embarrassment). But "blushing has remedial properties" (Dijk, de Jong, and Peters 2009, p. 290). Catherine's second blush, and at the very least her saying out loud, "I do not know what I was thinking of," can also be explained as a conscious choice to "undo" her first blush.

INSTINCTS

Another competing model is that people's actions are determined by drives or instincts, not choices. For example, when Henry Crawford is introduced to Fanny's father, Mr. Price becomes "a very different man.... His manners now, though not polished, were more than passable.... Such was his instinctive compliment to the good manners of Mr. Crawford" (MP, p. 467). Mr. Price does not consciously choose to stop swearing for example, and might not even be aware that he is changing his behavior; rather, he changes "instinctively" to meet the demands of the situation. Austen provides other examples relating to politeness and social presentation. When Mr. Darcy hands Elizabeth a letter answering the charges she leveled in refusing his proposal, "she instinctively took" it, and when he greets her later at Pemberley, Elizabeth, embarrassed, "instinctively turned away" (PP, pp. 218, 278). Hiding can also be instinctive. When Sir Thomas returns, Mrs. Norris, with "instinctive caution... whisked away Mr. Rushworth's pink satin cloak," his costume for the play (MP, p. 210), and when Catherine is exploring the hidden parts of Northanger Abbey, "[a]n attempt at concealment had been her first instinctive movement on perceiving" General Tilney (NA, p. 197). According to Hodgson (2010, p. 3), "[p]urged from the Anglophone social sciences in the interwar years, the concept of instinct has now returned.... [W]e use the term broadly to describe any biologically inherited reflex, feeling or disposition that can be triggered by specific cues." Cosmides and Tooby (1994, p. 64) argue that instincts "develop without any conscious effort and in the absence of any formal instruction."

But Austen is skeptical of instinct, especially for more important matters. Fanny is invited by Mrs. Grant to dine at the parsonage, and Edmund tells Fanny that he has successfully convinced Lady Bertram to allow her to go. " 'Thank you, I am *so* glad,' was Fanny's instinctive reply; though when she had turned from him and shut the door, she could not help feeling, 'And yet why should I be glad? for am I not certain of seeing or hearing something there to pain me?' " (MP, p. 256). Fanny instinctively thanks Edmund even though at dinner she will painfully observe Edmund and her rival, Mary Crawford, together. The maternal instinct is presumably one of the strongest, but when Fanny returns home to her mother after several years, she is disappointed at how little attention she receives: "the instinct of nature was soon satisfied, and Mrs. Price's attachment had no other source.... [S]he had neither leisure nor affection to bestow on Fanny" (MP, p. 450). At Fanny's home in Portsmouth, her father asks whether the newspaper story he reads about a Mrs. R running off with a Mr. C could be about Fanny's cousin, and Fanny denies it "from the instinctive wish of delaying shame," which is useless since she already knows that it must be true, given the letter she has just received from Mary Crawford imploring her not to believe rumors about her brother (MP, p. 509). Before Emma meets Frank Churchill, "her own imagination had already given her such instinctive knowledge," but of course most of what she thinks about Frank Churchill turns out to be wrong (E, p. 131). Walking together down the street in Bath, Anne worries about how Lady Russell will react when she sees Captain Wentworth for the first time in several years. As Captain Wentworth approaches among a crowd, Anne "looked instinctively at Lady Russell...from time to time, anxiously; and...she was yet perfectly conscious of Lady Russell's eyes being turned exactly in the direction for him, of her being in short intently observing him" (P, p. 194). But Lady Russell was looking intently only at some window curtains, and Anne's instinct fails her: "in all this waste of foresight and caution, she...lost the right moment for seeing whether he saw them" (P, p. 195). More positively, when Louisa Musgrove falls unconscious at Lyme, "Anne, attending with all the strength and zeal, and thought, which instinct supplied, to Henrietta, still tried, at intervals, to suggest comfort to the others" (P, p. 119). Here instinct helps not by directly causing any particular action but by supplying capacity: strength, zeal, and most of all, thought.

HABITS

Habits can also explain a person's actions: "the term 'habit' generally denominates a more or less self-actuating disposition or tendency to engage in a previously adopted or acquired form of action" (Camic 1986, p. 1044). Habits are "moulded by environmental circumstances and transmitted culturally rather than biologically" (Hodgson 2010, p. 4).

Austen acknowledges that habits can affect behavior, but she does not like them. Most habits in Austen's novels are bad. The most commonly mentioned habits are habits of overconsumption, selfishness, and carelessness. To explain his marriage to the wealthy Miss Grey, Willoughby tells Elinor that "I had always been expensive, always in the habit of associating with people of better income than myself," and Elinor reflects on his "habits of idleness, dissipation, and luxury" (SS, pp. 363, 375). Similarly, Colonel Fitzwilliam tells Elizabeth that as a younger son, "[o]ur habits of expence make us too dependant, and there are not many in my rank of life who can afford to marry without some attention to money" (PP, p. 206). Edmund Bertram worries whether Mary Crawford's habits will keep her from accepting his proposal: "I have no jealousy of any individual....It is the habits of wealth that I fear" (MP, p. 489). Caroline Bingley and her sister are "in the habit of spending more than they ought" (PP, p. 16), Mr. Yates "had not much to recommend him beyond habits of fashion and expense" (MP, p. 142), and Tom Bertram recovers from his illness "without regaining the thoughtlessness and selfishness of his previous habits" (MP, p. 534). Mr. Smith, the husband of Anne's impoverished friend Mrs. Smith, had "careless habits" and was financially ruined by his friendship with Mr. Elliot, who had "bad habits" and was "careless in all serious matters" (P, pp. 226, 174). Even Mr. Woodhouse's endearing but annoying obsession with his own health is due to "his habits of gentle selfishness" (E, p. 6), and similarly Anne's hypochondriac sister Mary Musgrove is "always in the habit of claiming Anne when any thing was the matter" (P, p. 35).

Some habits are good. Admiral and Mrs. Croft "brought with them their country habit of being almost always together" (P, p. 182). Louisa and Henrietta Musgrove are also always together, and Anne admires "the sort of necessity which the family-habits seemed to produce, of every thing being to be communicated, and everything being to be done together" (P, p. 89); the closest thing Anne and her sister Mary have is the "habit of running in and out of each other's house at all hours" (P, p. 39). Sir Thomas has "correctly punctual habits" (MP, p. 259), Emma's sister Isabella and her father share "a strong habit of regard for every old acquaintance" (E, p. 100), and Emma appreciates "the strong domestic habits, the all-sufficiency of home to himself" of her brother-in-law Mr. John Knightley (E, p. 104). Henry Tilney jokes that as a young woman Catherine must have a "delightful habit of journalizing" (NA, p. 19). But these good habits have little of the power of the bad ones, and do not help people nearly to the degree that the bad ones drag them down, into loveless marriages, illness, or financial ruin.

The most relevant habits, those that directly affect the life courses of our heroines, are painful encumbrances that must be thrown off. After Anne and Captain Wentworth finally understand each other, Captain Wentworth admits how discouraged he was by his memories of being rejected eight years earlier. Anne says that he should have realized how much older she is now and how different the current situation is, but Captain Wentworth explains, "I could not derive benefit from the late knowledge I had acquired of your character. I could not bring it into play: it was overwhelmed, buried, lost in those earlier feelings which I had been smarting under year after year.... The force of habit was to be added" (P, p. 266). When Tom Bertram tells Fanny that they need her services (to play Cottager's Wife), she "was up in a moment, expecting some errand, for the habit of employing her in that way was not yet overcome, in spite of all that Edmund could do" (MP, p. 171). After receiving Henry Crawford's proposal, Fanny strenuously avoids Mary Crawford, but when Mary asks to speak with her alone, Fanny's "habits of ready submission ... made her almost instantly rise and lead the way out of the room. She did it with wretched feelings, but it was inevitable" (MP, p. 412). Habits make Fanny submissive and miserable and must be overcome.

Habits can lead to moral depravity and even borderline evil. After Henry Crawford runs off with the married Maria, Edmund is stunned that Mary thinks their action only indiscreet, not immoral, and tells

Mary so. Edmund glimpses a brief moment of moral awareness, only to watch her pulled back down by her habits: "I imagined I saw a mixture of many feelings—a great, though short struggle—half a wish of yielding to truths, half a sense of shame—but habit, habit carried it.... It was a sort of laugh, as she answered, 'A pretty good lecture, upon my word. Was it part of your last sermon?'" (MP, p. 530). Henry Crawford proposed to Fanny immediately after telling her that he helped get her brother William promoted, a quid pro quo that disgusts Fanny: "such were his habits, that he could do nothing without a mixture of evil" (MP, p. 350).

For Austen, habits and rational choice are not necessarily opposed: sometimes, habits can positively provide the resolution necessary to make a choice. Emma thinks that Frank Churchill cannot easily come away to visit them because of "habits of early obedience" to the Churchill family, but Mr. Knightley maintains, "It ought to have been an habit with him by this time, of following his duty, instead of consulting expediency" (E, p. 159). Similarly, when Harriet continues to pine over Mr. Elton, even as he is about to marry another, Emma wishes that Harriet would have greater "habit of self-command" (E, p. 288).

Twice Austen allows both habit and rational choice to be relevant in an explanation, and both times rational choice is found more important. Edmund goes away for three weeks, and "[w]hat was tranquillity and comfort to Fanny was tediousness and vexation to Mary. Something arose from difference of disposition and habit—one so easily satisfied, the other so unused to endure; but still more might be imputed to difference of circumstances. In some points of interest they were exactly opposed to each other" (MP, p. 331). Fanny is grateful for any delay of progress between Edmund and Mary, while Mary is worried that she had so openly mocked the clergy (Edmund's chosen profession) and that Edmund might meet other attractive women. Habit explains some of the difference between Fanny's and Mary's feelings, but the difference in their preferences, their "points of interest," explains more.

Similarly, Jane Fairfax tries to change the subject away from her morning visits to the post office by remarking on the reliability of the postal service: "So seldom that a letter, among the thousands that are constantly passing about the kingdom, is even carried wrong—and not one in a million, I suppose, actually lost!" Mr. John Knightley agrees: "[T]he clerks grow expert from habit.—They must begin with some quickness of sight and hand, and exercise improves them. If you want any farther explanation...they are paid for it. That is the key to a great deal of capacity. The public pays and must be served well" (E, p. 320). Habit is important for the clerks' reliability, but the underlying explanation is that they want to get paid. As Guenter Treitel (personal correspondence) points out, Austen must establish the reliability of the postal service so that later Jane Fairfax, having written Frank Churchill and receiving no response, can therefore conclude that he does not care, because his reply could not possibly have been delayed or lost in the mail. But nothing in the plot turns on the specific explanation for postal service reliability. Perhaps Mr. John Knightley's statement that the public must be served well is about government accountability; according to Sutherland and Le Faye (2005, p. 203), this remark indicates his interest in a political career. But this does not explain why he explicitly compares the explanation of being paid with the explanation of habit, or why he talks about explanations at all. Mr. John Knightley's favoring rational choice over habit betrays Austen's theoretical stance.

Finally, Edmund thinks that Fanny refuses Henry Crawford because she is "of all human creatures the one, over whom habit had most power, and novelty least: and that the very circumstance of the novelty of Crawford's addresses was against him" (MP, p. 409). But Edmund is completely wrong. Fanny does not like Henry Crawford. Fanny refuses him because of her preferences, not her habits.

RULES

Another competing model is that people base their actions on rules or principles, such as "when traveling, never pass up the opportunity to use the bathroom." One reason for using a rule is to avoid the cognitive difficulties of making a choice, such as estimating when one will get to the next bathroom stop. Rules "allow decision makers to process information in a less effortful manner" (Shah and Oppenheimer 2008, p. 207). For example, Henry and Mary Crawford offer to pick up Fanny from Portsmouth and return her to Mansfield Park, but Fanny knows that the terms of her conveyance are up to her uncle, Sir Thomas. Fanny weighs the pros and cons: returning to Mansfield "was an image of the greatest felicity—but it would have been a material drawback, to be owing such felicity to persons in whose feelings and conduct, at the present moment, she saw so much to condemn" (MP, p. 504). Fanny is not even sure that she is capable of "judging impartially," given her aversion to any scheme that allows Mary Crawford to come to Mansfield and thereby see Edmund again (MP, p. 504). But Fanny has a rule: "Happily, however, she was not left to weigh and decide between opposite inclinations and doubtful notions of right; there was no occasion to determine, whether she ought to keep Edmund and Mary asunder or not. She had a rule to apply to, which settled every thing. Her awe of her uncle, and her dread of taking a liberty with him, made it instantly plain to her, what she had to do. She must absolutely decline the proposal" (MP, p. 505). The rule makes things easy; Fanny does not have to weigh her own preferences and choose herself.

A rule or principle can be a moral imperative, a requirement of proper conduct, or simply a useful guideline. Elinor suffers alone with the secret of Lucy Steele's engagement to Edward Ferrars because she made a promise: "she was firmly resolved to act by her as every principle of honour and honesty directed" (SS, p. 163). Keeping a promise is a moral imperative. Once the engagement becomes public, Elinor does her best not to talk about it: "Elinor avoided it upon principle, as tending to fix still more upon her thoughts, by the too warm, too positive assurances of Marianne, that belief of Edward's continued affection for herself which she rather wished to do away" (SS, p. 306). By avoiding the subject on principle, Elinor protects her heart from Marianne's suggestions that Edward really loves her instead. This principle is not a moral imperative, just a good idea. Fanny's rule of leaving her return up to her uncle is somewhere between the two: respecting her uncle is virtuous but not quite a moral imperative, and it is also just a good idea not to cross him. Similarly, Catherine Morland's parents want to approve her marriage to Henry Tilney, "but their principles were steady, and while his parent so expressly forbad the connection, they could not allow themselves to encourage it" (NA, p. 258). Waiting for General Tilney to approve before giving their own approval is not a moral imperative, but is consistent with proper conduct and good for the two families' future harmony.

Principles and rules are important to Austen, and people are often condemned for having poor or nonexisting ones. For example, Edmund Bertram tells Fanny that Mary Crawford's faults "are faults of principle, Fanny, of blunted delicacy and a corrupted, vitiated mind" (MP, p. 528). Principle is often understood by Austen as governing and steadying a person's motivations; for example, Sir Thomas, contemplating his poor parenting of his daughters, recognizes that "principle, active principle, had been wanting, that they had never been properly taught to govern their inclinations and tempers" (MP, p. 536).

But Austen notes that the relationship between principles and choice is not so simple. For example, Mrs. Norris's miserliness, an "infatuating principle" in which she takes pride, began because of her husband's low income, but with no children of her own, "what was begun as a matter of prudence, soon grew into a matter of choice" (MP, p. 9). A principle can govern choice, but one can also choose one's principles.

Also, a principle can heighten as well as moderate preferences. When Sir Thomas notices that Fanny's room does not have a fire, he explains that Mrs. Norris espouses "moderation in every thing.... The principle was good in itself, but it may have been, and I believe *has been* carried too far in your case.... Though their caution may prove eventually unnecessary, it was kindly meant; and of this you may be assured, that every advantage of affluence will be doubled by the little privations and restrictions that may have been imposed" (MP, p. 361). Here a principle increases Fanny's consumer satisfaction later in life: a satisfaction doubler, not a satisfaction governor. Fanny vows to "overcome all that was excessive, all that bordered on selfishness in her affection for Edmund. . . . She would endeavour to be rational, and to deserve the right of judging of Miss Crawford's character and the privilege of true solicitude for him by a sound intellect and an honest heart. She had all the heroism of principle." By following principles, Fanny wants to be unselfish and fair to Edmund and also protect her own heart. But "after making all these good resolutions on the side of self-government, she seized the scrap of paper on which Edmund had begun writing to her, as a treasure beyond all her hopes. . . . The enthusiasm of a woman's love is even beyond the biographer's" (MP, pp. 307–8). Like a person who after exercising feels entitled to extra dessert (Cloud 2009), Fanny's principles do not govern but intensify her desire.

Finally, Austen encourages skepticism by noting that a rule or principle is often just a cover for going ahead and doing what you want anyhow. For example, Mrs. Musgrove, Anne's sister's mother-in-law, says, "I make a rule of never interfering in any of my daughter-in-law's concerns, for I know it would not do; but I shall tell *you,* Miss Anne, because you may be able to set things to rights, that I have no very good opinion of Mrs. Charles's nursery-maid" (P, p. 49). After strenuous attempts to persuade Fanny to accept Henry Crawford, "Sir Thomas resolved to abstain from all farther importunity with his niece, and to shew no open interference. . . . Accordingly, on this principle Sir Thomas took the first opportunity of saying to her, . . . 'My dear. . . . You cannot suppose me capable of trying to persuade you to marry against your inclinations' " (MP, pp. 380–81). But Sir Thomas soon sends Fanny to Portsmouth to make her change her mind. Talking with Catherine about men, Isabella Thorpe declares, "I make it a rule never to mind what they say. They are very often amazingly impertinent," only later to welcome Captain Tilney's impertinent flirting (NA, p. 35). General Tilney says, "It is a rule with me, Miss Morland, never to give offence to any of my neighbours, if a small sacrifice of time and attention can prevent it," not long before kicking Catherine out (NA, p. 216). Emma tells Harriet, "I lay it down as a general rule, Harriet, that if a woman *doubts* as to whether she should accept a man or not, she certainly ought to refuse him," but of course this is merely a ploy to make Harriet reject Mr. Martin's proposal (E, p. 55). Elinor tells Marianne that for Willoughby, "[h]is own enjoyment, or his own ease, was, in every particular, his ruling principle" (SS, p. 398), and similarly Anne concludes that "Mr. Elliot...has never had any better principle to guide him than selfishness" (P, p. 225).

SOCIAL FACTORS

Social factors such as envy, pride, distinction, duty, honor, and decorum are another set of competing considerations. These factors are sometimes considered outside the scope of rational choice explanations; for example, Cramer (2002, p. 1846) writes that "rational choice theories...sack the social." Austen observes, however, that social factors can be brought in as part of a person's preferences. For example, Mrs. Price, Fanny's mother, had married badly; after being condemned by her sister Mrs. Norris, she no longer spoke to her sisters. "By the end of eleven years, however, Mrs. Price could no longer afford to cherish pride or resentment, or to lose one connection that might possibly assist her" (MP, p. 5). Mrs. Price's pride had kept her from choosing to contact her sisters, but eventually a low income, disabled husband, and ninth child provided enough motivation to outweigh it. In another example, after John Thorpe tries to trick Catherine into joining the ride to Blaize Castle, "her own gratification...might have been ensured in some degree by the excursion itself," but Catherine chooses to keep her engagement with the Tilneys because of "what was due to others, and to her own character in their opinion" (NA, p. 102). In other words, social factors such as pride, obligation, and how people think of you can be factors in your preferences commensurate with other factors such as financial need and gratification.

Austen often condemns these social factors. Lady Catherine tries to force Elizabeth to promise not to marry Mr. Darcy by threatening, "[H]onour, decorum, prudence, nay, interest, forbid it. Yes, Miss Bennet, interest; for do not expect to be noticed by his family or friends, if you wilfully act against the inclinations of all. You will be censured, slighted, and despised, by everyone connected with him. Your alliance will be a disgrace; your name will never even be mentioned by any of us" (PP, p. 394). Lady Catherine admits that it all boils down to interest and that honor and decorum are mere covers for hard-core social shunning, "relational aggression" (Crick and Grotpeter 1995; see also Bender 2012, p. 190). Taylor (2006, p. xiv) argues that game theory, by thinking of social norms mainly in terms of sanctions like shunning and censure, ignores "the essential characteristic of a norm—its normativity," its "moral motivation." Austen argues that one person's normativity is another person's scam.

When Fanny refuses Henry Crawford's proposal, people press her with every social factor they can think of. Mary Crawford appeals to social distinction, that Fanny has bested the many other women interested in Henry: "Oh! the envyings and heart-burnings of dozens and dozens! the wonder, the incredulity that will be felt at hearing what you have done!... [T]he glory of fixing one who has been shot at by so many" (MP, pp. 416–19). Mary is also not above appealing to conformity: "It is you only, you, insensible Fanny, who can think of him with any thing like indifference" (MP, p. 418). As mentioned in chapter 5, Henry Crawford appeals to reciprocity by securing William Price's promotion, and Lady Bertram appeals to Fanny's duty. Sir Thomas lectures, "You do not owe me the duty of a child. But, Fanny, if your heart can acquit you of *ingratitude,*" and Fanny bursts into tears (MP, p. 368). Against all these social factors, Fanny's resolve is heroic. Even the independent Emma, for example, hosts a dinner for the despised Eltons only because of social sanctions: "They must not do less than others, or she should be exposed to odious suspicions, and imagined capable of pitiful resentment" (E, p. 314).

Social norms are often considered a necessary corrective to unbridled selfishness, but it is easy to be in favor of social norms when they are not stacked against you. Shouldn't Fanny be able to make a "selfish," "individualistic" choice about whether or whom to marry? The idea that social factors are a bulwark against corrosive self-interest might be considered an affectation, even weapon, of the privileged.

IDEOLOGY

It is often argued that social factors do not simply affect a person's preferences, but rather create the entire ideological environment in which she makes a decision: "action is always socially situated and cannot be explained by reference to individual motives alone" (Granovetter 1990, pp. 95–96). For example, Fanny's decision of whether to accept Henry Crawford is not just about the costs and benefits of going against her uncle's wishes; it also must be understood within the context of Fanny being a subservient niece who has lived in his house for more than eight years. "[T]he slightest 'reaction' of an individual to another is pregnant with the whole history of these persons and of their relationship" (Bourdieu and Wacquant 1992, p. 124). Your ideological environment can even influence what you think your own interests are, as in "false consciousness." Scott (1990, p. 72) describes a "thick" version and "thin" version of how ideology affects a person's values: "the thick version claims that a dominant ideology works its magic by persuading subordinate groups to believe actively in the values that explain and justify their own subordination.... The thin theory... maintains only that the dominant ideology achieves compliance by convincing subordinate groups that the social order in which they live is natural and inevitable. The thick theory claims consent; the thin theory settles for resignation."

The closest Austen comes to addressing this issue is when Elinor, Marianne, and Edward Ferrars discuss how best to form opinions of other people. Elinor often finds "people so much more gay or grave, or ingenious or stupid than they really are, and I can hardly tell why, or in what the deception originated. Sometimes one is guided by what they say of themselves, and very frequently by what other people say of them, without giving oneself time to deliberate and judge." Marianne teases, "But I thought it was right, Elinor... to be guided wholly by the opinion of other people. I thought our judgments were given us merely to be subservient to those of our neighbours. This has always been your doctrine, I am sure." Elinor then makes her doctrine clear: "No, Marianne, never. My doctrine has never aimed at the subjection of the understanding. All I have ever attempted to influence has been the behaviour.... I am guilty, I confess, of having often wished you to treat our acquaintance in general with greater attention; but when have I advised you to adopt their sentiments or to conform to their judgment in serious matters?" (SS, p. 108).

This light conversation is Austen's defense of independent thought and rejection of the thick theory of ideology (see also Waldron 1999, p. 67). Others can affect your behavior through norms of what is socially expected, but you cannot allow them to affect your judgment or thought processes. Marianne later applies this doctrine when Mrs. Dashwood tells her to stay longer in London even though she longs to return home: "Marianne had promised to be guided by her mother's opinion, and she submitted to it therefore without opposition, though it proved perfectly different from what she wished and expected, though she felt it to be entirely wrong, formed on mistaken grounds" (SS, p. 243).

According to Scott, there is little empirical support for either theory of ideology, thick or thin, and the only reason we seem to observe false consciousness is because oppressed people strategically keep their heretical thoughts to themselves until they find the right opening (see also Kelley 1993). Fanny is well aware of the social context that makes a marriage between herself and Edmund inconceivable because of her disadvantages of family and fortune, but she does not allow these strictures to limit her thought processes or change what she knows she wants, even though she knows that they should: "To think of him as Miss Crawford might be justified in thinking, would in her be insanity. To her, he could be nothing under any circumstances—nothing dearer than a friend. Why did such an idea occur to her even enough to be reprobated and forbidden? It ought not to have touched on the confines of her imagination" (MP, p. 307). Ideology is no match for a woman's love. Once Mary Crawford is out of the picture, Fanny has her opening.

INTOXICATION

People are not always in control of themselves and can do random and unpredictable things. This might be considered another competing model: "the possibility that the phenomena studied in the social and behavioral sciences are essentially unpredictable and indeterminate" (Cziko 1989, p. 17). But even when one of Austen's characters makes a mistake, it is almost always consistent with her personality and values, such as Emma's glib insult toward Miss Bates and Catherine's suspicion that General Tilney murdered his wife. When Marianne bursts into tears on Elinor's shoulder, "Colonel Brandon rose up and went to them without knowing what he did"; his action is not a conscious choice but is consistent with his devotion to Marianne (SS, p. 269).

Even under the influence of alcohol, Austen's characters behave intelligently and with coherent purpose. When Mr. Elton proposes to Emma in the carriage, he "had only drunk wine enough to elevate his spirits, not at all to confuse his intellects" (E, p. 141). When Willoughby arrives unexpectedly to apologize to Marianne, Elinor thinks he is drunk, but "[t]he steadiness of his manner, and the intelligence of his eye as he spoke, convinc[es] Elinor, that whatever other unpardonable folly might bring him to Cleveland, he was not brought there by intoxication" (SS, p. 361).

The closest example of someone doing something completely unaccountable is when Elizabeth dances at the Netherfield ball: "Mr. Darcy...took her so much by surprise in his application for her hand, that, without knowing what she did, she accepted him" (PP, p. 101). Even in this case, Elizabeth has a moment to reconsider before the dance starts, but consciously chooses to go ahead, prodded by Charlotte Lucas, who sensibly advises her "not to be a simpleton and allow her fancy for Wickham to make her appear unpleasant in the eyes of a man of ten times his consequence" (PP, p. 102).

CONSTRAINTS

Finally, Austen's single young women, who have almost no economic independence and no career path other than wife or governess, might seem so constrained that they hardly have choices at all. In other words, another model to consider is that a person's constraints, not her choices, best explain her actions. Duesenberry (1960, p. 233, quoted in Abbott 2004, p. 49) writes, "Economics is all about how people make choices. Sociology is all about why they don't have any choices to make."

Austen's most direct response to this is a passage in which Fanny, having received Henry Crawford's proposal, dreads speaking alone with his ally, Mary Crawford. Fanny therefore "absented herself as little as possible from Lady Bertram, kept away from the east room, and took no solitary walk in the shrubbery, in her caution to avoid any sudden attack" (MP, p. 412). Fanny's plan works for a while, but "Miss Crawford was not the slave of opportunity. She was determined to see Fanny alone, and therefore said to her tolerably soon, in a low voice, 'I must speak to you for a few minutes somewhere'" (MP, p. 412). No person must remain a slave to given opportunities.

Fanny, who talks with Mary Crawford because "[d]enial was impossible," is admittedly less independent than Mary, but even Fanny can execute the strategy of avoiding being alone, which works to some degree (MP, p. 412). Perhaps the entire point of *Mansfield Park* is that even someone as dependent as Fanny can learn to think strategically and eke out space to maneuver. Some of Fanny's stratagems are perhaps trivial, such as staying close to Lady Bertram, but some are not. When Mary Crawford wants to talk to her, Fanny avoids the shrubbery, but when Edmund declares that "I will take the first opportunity of speaking to her alone," his father informs him that Fanny is "at that very time walking alone in the shrubbery" (MP, p. 400).

For that matter, Austen observes how social constraints can make you learn strategic thinking more quickly; Nelles (2006, p. 127) notes that in Austen's novels, the "need to develop telepathy" is due to "women's enforced passivity and immobility." As mentioned in chapter 5, Fanny's first manipulation is persuading the upset Mr. Rushworth, who has returned with the key to his gate, to go and join Maria Bertram and Henry Crawford even though they already left without him. Fanny acts because social decorum, everyone getting along, requires it; had she not been constrained to act as a young woman is expected to act, she would not have had to figure out that the best way to make Mr. Rushworth move is to flatter him.

Similarly, Elinor is socially constrained by her promise not to reveal Lucy's secret engagement with Edward Ferrars. After the engagement is finally made public, Elinor explains to Marianne, "My promise to Lucy, obliged me to be secret. I owed it to her, therefore, to avoid giving any hint of the truth; and I owed it to my family and friends, not to create in them a solicitude about me, which it could not be in my power to satisfy....I have very often wished to undeceive yourself and my mother...and once or twice I have attempted it;—but without betraying my trust, I never could have convinced you. . . . The composure of mind with which I have brought myself at present to consider the matter, the consolation that I have been willing to admit, have been the effect of constant and painful exertion;—they did not spring up of themselves" (SS, pp. 297–99). Elinor, considering what her mother and sister would do if they knew, tries to keep them from sharing her unhappiness. Elinor tries to solve the difficult puzzle of how to tell them without directly betraying Lucy, but in the end, reasoning strategically, concludes that without breaking her promise outright they would not believe her. Constrained to silence, Elinor must call upon her deepest personal reserves of effort and exertion.

CHAPTER EIGHT

Austen on What Strategic Thinking Is Not

AUSTEN CAREFULLY distinguishes strategic thinking from concepts possibly confused with it: selfishness, moralistic notions of what a person "should" do, economistic values, and winning inconsequential games. Like any social theorist, Austen seeks conceptual clarity. But she also wants to make particularly clear that she is not advocating selfishness or money-centrism or one-upmanship or anything as vulgar as telling young women "how to behave." Strategic thinking should not be confused with a set of hackneyed prescriptions.

STRATEGIC THINKING IS NOT SELFISH

For Austen, strategic thinking is not equivalent to selfishness. Of course some people exemplify both strategicness and selfishness, such as Willoughby and Lucy Steele. But one can be strategic with benevolent intentions: for example, Sir Thomas sends Fanny to dreary Portsmouth to make her better appreciate the material comforts of accepting Henry Crawford's marriage proposal. Sir Thomas is "delighted with his own sagacity" in this plan and genuinely believes, not without reason, that he is acting to advance Fanny's happiness (MP, p. 449). One can also be selfish and a bad strategist: "cold hearted selfishness" and "a general want of understanding" are combined in Fanny Dashwood, whose mistakes include keeping Elinor away from her brother Edward Ferrars by bringing in the more dangerous Lucy Steele (SS, p. 261).

Once she marries Robert Ferrars, Lucy is described as having "selfish sagacity," not just "sagacity," in flattering her way back into the favor of her mother-in-law Mrs. Ferrars (SS, p. 426). Sagacity—in other words, strategic thinking—does not by itself imply selfishness. Similarly, Elinor calls Lucy "illiterate, artful, and selfish" (SS, p. 160), and Anne Elliot observes to Mrs. Smith that "[t]he manoeuvres of selfishness and duplicity must ever be revolting" (P, p. 224). Even though the terms "artful" and "manoeuvre" have a connotation of craftiness, they do not by themselves indicate selfishness. At Bath, Mrs. Clay and Mr. Elliot both hover around Sir Walter's household, with Mrs. Clay hoping to marry Sir Walter and Mr. Elliot hoping to prevent it to make sure he will inherit Sir Walter's estate. Mr. Elliot's greater strategic sophistication implies a more complicated, not a greater, selfishness: "Mrs. Clay's selfishness was not so complicate nor so revolting as his; and Anne would have compounded for the marriage at once, with all its evils, to be clear of Mr. Elliot's subtleties, in endeavouring to prevent it" (P, p. 233). If anything, as discussed in chapter 7, Austen associates selfishness not with strategic thinking but with habit, as in Dr. Grant's "very faulty habit of self-indulgence" and Henry Crawford's "own habits of selfish indulgence" (MP, pp. 130, 275).

Austen makes clear that having goals and strategically working toward them does not make one selfish or self-interested. The goal of Austen's heroines is to marry, but sincere affection is different from self-interest. This is how Elinor evaluates Lucy's engagement with Edward Ferrars: "he had not even the chance of being tolerably happy in marriage, which sincere affection on *her* side would have given, for self-interest alone could induce a woman to keep a man to an engagement, of which she seemed so thoroughly aware that he was weary" (SS, p. 173). Sincere affection requires mutuality.

Saying that making one's own choices is equivalent to selfishness can simply be a method of intimidation. Mrs. Norris complains that Fanny "likes to go her own way to work; she does not like to be dictated to; she takes her own independent walk whenever she can; she certainly has a little spirit of secrecy, and independence, and nonsense" (MP, p. 373). Mrs. Norris cannot stand Fanny making independent choices, even over matters as trivial as when and where to take a walk. Hearing this, Sir Thomas "thought nothing could be more unjust, though he had been so lately expressing the same sentiments himself" in pushing Fanny to accept Henry Crawford (MP, p. 373). When Elizabeth refuses to promise never to enter into an engagement with Mr. Darcy, Lady Catherine cries, "Unfeeling, selfish girl! Do you not consider that a connection with you, must disgrace him in the eyes of everybody?" (PP, p. 396). For Mrs. Norris and Lady Catherine, calling a young woman independent or selfish is just another way to keep her from making her own choices.

STRATEGIC THINKING IS NOT MORALISTIC

One can easily confuse what a person wants with what she "should" want; for example, it is tempting to say that a person who smokes five packs of cigarettes a day cannot be making a rational choice. But Austen makes this distinction clear. When Mrs. Weston suggests that Mr. Knightley might propose to Jane Fairfax, Emma gets flustered: "Dear Mrs. Weston, how could you think of such a thing?—Mr. Knightley!— Mr. Knightley must not marry!—You would not have little Henry cut out from Donwell?—Oh! no, no, Henry must have Donwell. I cannot at all consent to Mr. Knightley's marrying; and I am sure it is not at all likely" (E, p. 242). Emma speaks for little Henry, Mr. Knightley's nephew (and heir if he remains unmarried), but of course the question is not what Mr. Knightley should do but what he will do, as Mrs. Weston points out: "the question is not, whether it would be a bad connexion for him, but whether he wishes it; and I think he does" (E, p. 244).

Strategic thinking is not about moralistic maxims of what one should do. Mary Bennet, Elizabeth's sister, is very bad at understanding the minds of others and likes to edify her sisters with "new observations of thread-bare morality" (PP, p. 67). For example, when Lydia has run away with Wickham and the family anxiously considers what should be done next, Mary quite unhelpfully proclaims, "[W]e may draw from it this useful lesson; that loss of virtue in a female is irretrievable—that one false step involves her in endless ruin," and in response Elizabeth "lifted up her eyes in amazement" (PP, p. 319). At a ball, Catherine Morland cannot accept the far preferable Henry Tilney because John Thorpe has already engaged her for the first dance. Thus "she deduced this useful lesson, that to go previously engaged to a ball, does not necessarily increase either the dignity or enjoyment of a young lady. From such a moralizing strain as this, she was suddenly roused by a touch on the shoulder, and turning round, perceived Mrs. Hughes directly behind her, attended by Miss Tilney and a gentleman. 'I beg your pardon, Miss Morland,' said she, 'for this liberty—but I cannot any how get to Miss Thorpe, and Mrs. Thorpe said she was sure you would not have the least objection to letting in this young lady by you' " (NA, pp. 50–51). On the verge of moralizing, Catherine much more usefully learns and benefits from Mrs. Hughes's simple strategic move of setting her up with Henry's sister.

STRATEGIC THINKING IS NOT ECONOMISTIC

Strategic thinking can also be confused with a variety of concepts related to "economy," such as thrift, materialism, and money-centrism. But Austen clearly distinguishes between economistic values and strategic thinking, especially through Mrs. Norris, who follows "a very strict line of economy," with "nothing to impede her frugality" (MP, p. 9). She wants to continue with the play *Lovers' Vows* regardless of the play's subject or whether Sir Thomas would approve, because "the preparations will be all so much money thrown away—and I am sure *that* would be a discredit to us all" (MP, p. 166). Sewing a curtain for the stage, she manages through meticulous planning to save a whopping three-fourths of a yard out of an entire bolt of green baize. When Sir Thomas returns and the play is called off, the curtain "went off with her to her cottage, where she happened to be particularly in want of green baize" (MP, p. 228). Mrs. Norris thinks of herself as a strategic sophisticate. But, as discussed in chapter 5, when she maneuvers to exclude Fanny from the trip to Sotherton, "more from partiality for her own scheme, because it *was* her own, than from any thing else," Edmund easily bests her, having already secured an invitation for Fanny (MP, p. 92). Mrs. Norris is proudest of Maria's marriage with Mr. Rushworth: "She took to herself all the credit of bringing Mr. Rushworth's admiration of Maria to any effect" (MP, p. 221). But the marriage's unsoundness should have been obvious from the start; at the wedding, Sir Thomas is anxious, but Mrs. Norris "was all joyous delight. . . . [N]o one would have supposed, from her confident triumph, that she . . . could have the smallest insight into the disposition of the niece who had been brought up under her eye" (MP, pp. 237–38).

The strategic sophomore John Dashwood also believes in the cash nexus, evaluating his sister Marianne's illness as an income reduction: "At her time of life, any thing of an illness destroys the bloom for ever! . . . I question whether Marianne *now*, will marry a man worth more than five or six hundred a-year, at the utmost" (SS, p. 258). When he hears that Colonel Brandon has offered a living to Edward Ferrars, he evaluates the offer not in terms of the Colonel's kindness but in the living's monetary value had he sold it instead: "[S]upposing the late incumbent to have been old and sickly, and likely to vacate it soon— he might have got I dare say—fourteen hundred pounds" (SS, p. 334).

Of course Austen's strategically skilled heroines do not ignore money altogether. Marianne believes that marriage cannot be "only a commercial exchange, in which each wished to be benefited at the expense of the other" (SS, p. 45), and that "money can only give happiness where there is nothing else to give it" (SS, p. 105), but famously states her basic needs as "[a]bout eighteen hundred or two thousand a-year," twice what Elinor thinks luxurious (SS, p. 105). When Elizabeth talks to Mrs. Gardiner about Wickham's attention to Miss King, who has just inherited a fortune from her grandfather, Elizabeth argues that one cannot pretend that cash is irrelevant: "Pray, my dear aunt, what is the difference in matrimonial affairs, between the mercenary and the prudent motive? Where does discretion end, and avarice begin? Last Christmas you were afraid of his marrying me, because it would be imprudent; and now, because he is trying to get a girl with only ten thousand pounds, you want to find out that he is mercenary" (PP, p. 173).

For Austen, strategic skills and money skills do not necessarily go together, but neither are they necessarily opposed. Mr. Shepherd, the lawyer of Sir Walter Elliot, has strategic skills in both the intimate sphere of coddling clients and the larger sphere of the market; Treitel (1984, p. 552) notes that "Mr. Shepherd is remarkably good at getting his own way." When Sir Walter has money problems, Mr. Shepherd is "skilful enough to dissuade" Sir Walter from relocating to London "and make Bath preferred" because it is cheaper (P, p. 15). When Sir Walter prohibits Mr. Shepherd from vulgarly advertising that Kellynch Hall is for rent, Mr. Shepherd, quite possibly having already communicated with Admiral Croft, advises that "peace will be turning all our rich Naval Officers ashore" and soon afterward tells Sir Walter that Admiral Croft has "accidentally hear[d] of the possibility of Kellynch Hall being to let" (P, p. 23). Mr. Shepherd skillfully finds a tenant while preserving his client's undeserved pride, and is likely emulated by his daughter Mrs. Clay, who "understood the art of pleasing" and "possessed, in an acute mind and assiduous pleasing manners, infinitely more dangerous attractions than any merely personal might have been" (P, pp. 17, 37).

STRATEGIC THINKING IS NOT ABOUT WINNING INCONSEQUENTIAL GAMES

Austen often features card games and parlor games, such as whist and backgammon. One might think that she would use these games to illustrate strategic thinking, as do other game theorists (such as Binmore 2007). Instead, Austen uses these games to illustrate the tendency of excessive decontextualization, of focusing in so closely on the inconsequential that one loses sight of the larger social context. Austen emphasizes that strategic thinking is about much more than the triviality of "winning."

Austen's characters who like card and board games are generally not good at strategic thinking in the social realm. "Lovers of games in the novels...are more often than not selfish, irresponsible, or empty-headed" (Duckworth 1975, p. 280). Mr. Collins, utterly misunderstanding the tastes of his audience, tries to entertain the Bennet daughters by reading from Fordyce's Sermons "with very monotonous solemnity" (PP, pp. 76–77). Interrupted by Lydia, who can't take it any longer, Mr. Collins takes refuge with Mr. Bennet in a game of backgammon. Mr. Hurst, Mr. Bingley's brother-in-law, "was an indolent man, who lived only to eat, drink, and play at cards" (PP, p. 38) and the evening after Fanny's first ball at Mansfield, a drowsy Lady Bertram asks Fanny to "Fetch the cards,—I feel so very stupid" (MP, p. 328). In contrast, Elizabeth would rather "amuse herself...with a book" than play cards (PP, p. 40), and when Lady Middleton proposes that they all play Casino,

Marianne declares, "I detest cards. I shall go to the piano-forte" (SS, p. 165). Elinor finds that Mrs. Jennings's parties, "formed only for cards, could have little to amuse her" (SS, p. 192).

When Austen's characters who are skilled in real-world strategic thinking do play cards, they play with an eye to a larger social context, with a more serious game in mind; they resist the decontextualization that close attention to an "artificial" game like whist requires (see also Silver [2009/2010] on Alexander Pope). Mr. Darcy and Mr. Bingley visit the Bennets, and Elizabeth doesn't know what Mr. Darcy is up to. After exchanging pleasantries about his sister, Mr. Darcy stands next to Elizabeth for several minutes in silence. Elizabeth is anxious to communicate, somehow, but her hopes "were overthrown, by seeing him fall a victim to her mother's rapacity for whist players. ... They were confined for the evening at different tables, and she had nothing to hope, but that his eyes were so often turned towards her side of the room, as to make him play as unsuccessfully as herself" (PP, p. 378). For Elizabeth and Mr. Darcy, whist is a superficial distraction from the real game of how they will reach some kind of understanding, through conversation or eye contact. When Willoughby courts Marianne, "[i]f their evenings at the park were concluded with cards, he cheated himself and all the rest of the party to get her a good hand"; Willoughby has a larger objective and card games are a mere vehicle (SS, p. 64). After dear Miss Taylor leaves to marry Mr. Weston, Emma tries to keep her father from lamenting her absence, and plans "by the help of backgammon, to get her father tolerably through the evening" (E, p. 7).

Card playing presents a strategic opportunity to get other people out of the way; according to Mullan (2012, p. 147), "the first purpose of games in [Austen's] novels is to divide and dispose her characters." Wickham sits between Lydia and Elizabeth as they play a card game called lottery tickets. Lydia "soon grew too much interested in the game, too eager in making bets and exclaiming after prizes, to have attention for anyone in particular. Allowing for the common demands of the game, Mr. Wickham was therefore at leisure to talk to Elizabeth, and she was very willing to hear him, though what she chiefly wished to hear she could not hope to be told, the history of his acquaintance with Mr. Darcy" (PP, p. 86). Here Lydia's immersion in the game is convenient for Elizabeth and Wickham, who put only customary effort in the game. When Lady Middleton proposes Casino, Elinor at first joins in the card game but quickly realizes that since Lucy Steele has already volunteered to work instead on a basket for Lady Middleton's daughter, she now has a chance to talk with Lucy alone about her secret engagement to Edward Ferrars. Elinor proposes, "[I]f I should happen to cut out, I may be of some use to Miss Lucy Steele, in rolling her papers for her, and there is so much still to be done to the basket, that it must be impossible I think for her [to] labour singly." Elinor and Lucy sit at a table to work while Marianne plays the pianoforte, which "was luckily so near them that Miss [Elinor] Dashwood now judged she might safely, under the shelter of its noise, introduce the interesting subject, without any risk of being heard at the card-table" (SS, p. 165). Here card playing is multiply useful, allowing Elinor and Lucy to talk alone and also causing Marianne to remove herself and in turn play music that covers their voices. When Mr. Bingley visits the Bennets, "Mrs. Bennet's invention was again at work to get every body away from him and her daughter" Jane. They all sit down to cards, and Elizabeth concludes that since they are all occupied, she can go off and write a letter in another room because "she could not be wanted to counteract her mother's schemes." But when Elizabeth returns to the drawing room, Jane and Mr. Bingley are alone together, the proposal has been made, and Elizabeth realizes "that her mother had been too ingenious for her" (PP, p. 383). While the others play cards, Elizabeth thinks she can cleverly peel off. But perhaps Mrs. Bennet sets up the card game anticipating Elizabeth's cleverness; realizing that Elizabeth is the greatest obstacle, Mrs. Bennet must remove her first. Elizabeth is neutralized not by playing cards but by her belief that everyone else is neutralized by playing cards.

Austen's longest episode of card playing is a game of speculation, in which players bid for higher cards with chips or counters. After dinner at Mansfield Park, there is a table for whist and a table for speculation, and Lady Bertram, equally ignorant of both, is directed by Sir Thomas toward speculation because "[h]e was a Whist player himself, and perhaps might feel that it would not much amuse him to have her for a partner" (MP, p. 278). Lady Bertram is not equal to participating even in the confined domain of card games, and cannot even choose which game to play; in contrast, Sir Thomas is able to think strategically not just within the game of whist but in the social context in which the games are embedded. Henry Crawford wants to toy with Fanny (he has not yet fallen hard for her), and offers to sit between Lady Bertram and Fanny to teach them both. In contrast to Willoughby, however, Henry actually wants to win the game and not just the girl. Henry is "pre-eminent in all the lively turns, quick resources, and playful impudence that could do honour to the game" and tries to get Fanny to be more competitive: "he had yet to inspirit her play, sharpen her avarice, and harden her heart, which, especially in any competition with William, was a work of some difficulty" (MP, p. 279). When her brother William bids for Fanny's queen, Henry makes sure that Fanny turns down the bid, and Edmund remarks, "Fanny had much rather it were William's....Poor Fanny! not allowed to cheat herself as she wishes!" (MP, p. 284).

As Edmund is aware, Fanny, who was "mistress of the rules of the game in three minutes," does not find the game itself very interesting and cannot separate it from the overall social context, in which she cares most about her dear brother (MP, p. 279).

While still playing, Henry suggests at great length that Edmund spend money to improve Thornton Lacey, his future home as a clergyman. Edmund replies, "I think the house and premises may be made comfortable...without any very heavy expense, and that must suffice me; and I hope may suffice all who care about me," thus frustrating Mary Crawford, who has greater monetary ambitions for Edmund (MP, p. 282). Mary overbids for another's jack to send a message, declaring aloud, "I will stake my last like a woman of spirit. No cold prudence for me. I am not born to sit still and do nothing. If I lose the game, it shall not be from not striving for it" (MP, p. 282). Mary wants the smaller card game to represent the larger game of getting Edmund to pursue a more lucrative career. She wins the hand but it "did not pay her for what she had given to secure it" (MP, p. 282). Austen's characters who are truly strategically skilled do not care about small games and know that they do not mean or represent anything.

Austen's Innovations

AUSTEN MAKES five particular advances in game theory. In this chapter I discuss four of them, leaving the last, her analysis of cluelessness, to chapter 12. First, Austen examines how two people form an intimate relationship by strategically acting in concert to manipulate a third. Second, she looks at how the relationship between a person's multiple selves can be more complex than a simple chain of command. Third, Austen considers how a person's preferences change, for example, when an alternative takes on a new social connotation. Fourth, Austen argues that true constancy is not the same as individual obstinacy but rather requires active faith and strategic thinking in cooperation.

PARTNERS IN STRATEGIC MANIPULATION

Marriage is, of course, the central goal in Austen's novels, and our heroines use their strategic skills toward this goal. But strategic thinking plays another crucial role: almost always, a couple's relationship is prefaced by working together to strategically manage, or at least monitor, other people. Strategic partnership is the truest foundation for marriage and intimacy. Strategic thinking does not assume atomistic individuals; indeed, Austen argues that strategic thinking in concert forms the basis of the closest human relationships.

Emma and Mr. Knightley might discover the strength of their affection only after the encroachment of rivals, but the ease and versatility of their strategic teamwork is evident early on. When Mr. Woodhouse can't stop talking about how his daughter Isabella and her husband Mr. John Knightley, Mr. Knightley's brother, should not have taken their vacation at South End because of the dangerous sea air, and for their own health should have visited him at Highbury instead, Emma changes the subject several times because she knows that her father's stubbornness risks an outright dispute. Despite her best efforts, her father persists and appeals to the authority of his apothecary friend, Mr. Perry. This is enough to make Mr. John Knightley exclaim, "Mr. Perry... would do as well to keep his opinion till it is asked for. Why does he make it any business of his, to wonder at what I do?" (E, p. 114). But Mr. Knightley jumps in, "with most ready interposition" and "immediate alertness," asking his brother for his opinion about moving one of the paths on his estate (E, pp. 114–15). Mr. Knightley thus demonstrates not only his situational awareness and strategic skills, but also his unspoken connection with Emma, their teamwork that needs no explicit planning. When the Woodhouses and Knightleys dine at Randalls, it begins to snow and the possibility looms of having to spend the night there or, even worse, getting stuck in the snow on the way home. Emma's sister Isabella wants to leave first with her husband so they can return to their children quickly, and thinks that her father and Emma should remain. But Emma, and her father even more so, would rather not spend the night, and the question of what should be done, and the seriousness of the snowfall, is discussed chaotically. "[W]hile the others were variously urging and recommending, Mr. Knightley and Emma settled it in a few brief sentences: thus—'Your father will not be easy; why do not you go?' 'I am ready, if the others are.' 'Shall I ring the bell?' 'Yes, do' " (E, p. 138). That the two of them would consult and come up with a plan for everyone to follow is taken for granted, and their telegraphic exchange illustrates the understanding they already share.

Accordingly, what flusters Emma and Mr. Knightley most is the possibility of being kicked off the team and replaced by another. Emma anticipates a strategically skilled Frank Churchill who "can adapt his conversation to the taste of every body, and has the power as well as the wish of being universally agreeable." Mr. Knightley responds by calling him a "practised politician, who is to read every body's character. . . . My dear Emma, your own good sense could not endure such a puppy when it came to the point" (E, pp. 161–62). Mr. Knightley's defensiveness is strong enough for Emma to notice: "To take a dislike to a young man, only because he appeared to be of a different disposition from himself, was unworthy the real liberality of mind which she was always used to acknowledge in him" (E, p. 162). Conversely, what gives Emma "severe pain" is Mr. Knightley's "superior degree of confidence towards Harriet" when he tells Harriet "that though he must go to London, it was very much against his inclination that he left home at all, which was much more (as Emma felt) than he had acknowledged to *her*" (E, p. 447). Mr. Knightley asks Harriet, not Emma, to understand his motivations and share in his decision.

Once their match is announced, the prospect of their perfect teamwork makes Emma especially happy: "High in the rank of her most serious and heartfelt felicities, was the reflection that all necessity of concealment from Mr. Knightley would soon be over. . . . She could now look forward to giving him that full and perfect confidence which her disposition was most ready to welcome as a duty" (E, p. 519). This seems an odd thing to be joyful about, since she could have been perfectly truthful with Mr. Knightley as just a friend and since marriage is no guarantee of perfect confidence anyhow. In her happiness Emma expresses how fundamental strategic partnership is to her idea of marriage. United, Emma and Mr. Knightley have more than enough strategic expertise to predict what will happen once they tell Mr. Weston of their engagement: "they had calculated from the time of its being known at Randall's, how soon it would be over Highbury; and were thinking of themselves, as the evening wonder in many a family circle, with great sagacity" (E, p. 511).

The strategic partnership between Edmund Bertram and Fanny starts cursorily: Edmund asks her opinion on whether he should act in *Lovers' Vows*, but his mind is actually already made up, because he wants to protect Mary Crawford from acting with a complete stranger. Edmund has been in the habit of talking with Fanny about Mary, frequently praising her and excusing her faults, to Fanny's chagrin. But Fanny's patient listening pays off when Edmund begins to have real doubts. Mary Crawford has already made fun of the clergy, Edmund's chosen profession, and joked about her intention to be very rich. Edmund confides, "She does not *think* evil, but she speaks it—speaks it in playfulness—and though I know it to be playfulness, it grieves me to the soul." Fanny, still not up to actually advising him, responds, "If you only want me as a listener, cousin, I will be as useful as I can; but I am not qualified for an adviser. Do not ask advice of *me*. I am not competent" (MP, p. 312). The one thing Fanny does recommend is that Edmund not tell her anything about Mary that he might regret later. Edmund assures her, "The time will never come.... You are the only being upon earth to whom I should say what I have said; but you have always known my opinion of her.... How many a time have we talked over her little errors! You need not fear me; I have almost given up every serious idea of her" (MP, p. 313). His reply elates Fanny not just because he is nearly giving up on Mary, but also because he firmly rejects her suggestion that the discourse between them might ever be limited. Fanny affirms, "I cannot be afraid of hearing any thing you wish to say. Do not check yourself. Tell me whatever you like" (MP, p. 314).

Fanny receives Henry Crawford's proposal while Edmund is away, and when he returns, "whom else had she to open her heart to? ... Fanny estranged from him, silent and reserved, was an unnatural state of things; a state which he must break through, and which he could easily learn to think she was wanting him to break through" (MP, pp. 399–400). Thus concluding that he should make the first move, Edmund goes to talk with her alone while she walks in the shrubbery. Later, "Miss Crawford's power was all returning," and Edmund plans to go to London to propose; "Fanny was the more affected from feeling it to be the last time in which Miss Crawford's name would ever be mentioned between them with any remains of liberty" (MP, pp. 404, 431). Again, why should being able to talk about a third person be so important, unless this confidence itself leads toward intimacy? When Henry Crawford runs off with the married Maria, and Edmund is stunned at how casually Mary treats their violation, the depth of his disappointment is measured in his reluctance to talk even with Fanny about it: "If he would now speak to her with the unreserve which had sometimes been too much for her before, it would be most consoling; but *that* she found was not to be. . . . Long, long would it be ere Miss Crawford's name passed his lips again, or she could hope for a renewal of such confidential intercourse as had been" (MP, p. 524). Now it is Fanny's turn to draw Edmund out, and by doing so at length she gains his heart.

Catherine Morland is alarmed about Captain Frederick Tilney's attentions toward Isabella, who is engaged to her brother James, and resolves to ask Henry Tilney to call his brother off. Henry doesn't take Catherine's worries seriously, and Catherine finally directly asks, "But what can your brother mean? If he knows her engagement, what can he mean by his behaviour? . . . [Y]ou must know your brother's heart" (NA, p. 155). Henry replies that he can only guess at it, and assures Catherine that his brother will soon leave Bath and forget about Isabella. After receiving her brother's letter saying that Isabella plans an engagement with Captain Tilney instead, Catherine strategizes with Henry and Eleanor Tilney about whether they should tell General Tilney their side of the story before Captain Tilney comes to ask for his father's consent. Henry and Eleanor predict that General Tilney will object to Isabella's lack of wealth and that Captain Tilney will not have the courage to ask his father in person anyhow. Catherine suggests that Henry still talk with General Tilney, "but he did not catch at the measure so eagerly as she had expected," and Henry takes an entirely passive approach: "[M]y father's hands need not be strengthened, and Frederick's confession of folly need not be forestalled. He must tell his own story" (NA, p. 215). When Catherine receives Isabella's letter saying that Captain Tilney has left for good, Catherine asks Henry, "Why should he pay her such attentions as to make her quarrel with my brother, and then fly off himself? . . . [F]or mischief's sake?" (NA, p. 225). Henry can only bow in assent and tell Catherine that she has difficulty understanding because her "mind is warped by an innate principle of general integrity, and therefore not accessible to the cool reasonings of family partiality, or a desire of revenge," thereby mollifying Catherine: "Catherine was complimented out of further bitterness" (NA, pp. 225–26).

Therefore, finally when his father ejects Catherine, Henry must travel to Catherine's home to apologize not just for his father but also for his own negligence of their implicit strategic partnership. Henry had consistently devalued Catherine's concerns about his brother and had never actually done anything to address them, even after his brother's encroachment proved her fears valid. Henry had not thought it necessary to represent Catherine's concerns to his father General Tilney, which at the very least would have shown his father that he is willing to speak up for her. He does not even share in Catherine's frustration when his brother's baseness is fully revealed, choosing to butter her up instead. "[B]y turning a deaf ear to" Catherine, Henry "is more graceful, but he is not essentially different from the General" (Johnson 1988, p. 38). Finally, Henry is not present to represent Catherine's interests when General Tilney kicks her out. Catherine's expulsion is thus not just the convulsion of a rude old man, but the final harsh result of repeated mild negligence, of choosing wit over exertion, of not taking a friendship seriously. A neglect this severe in a strategic partnership can only be repaired with a marriage proposal.

Captain Wentworth and Anne Elliot's strategic partnership is forged in the heat of medical emergency at Lyme; at first he sets off for the surgeon himself, but quickly agrees with Anne that Captain Benwick should go since he knows Lyme better. Captain Harville and his wife immediately jointly insist that Louisa be taken to their home to wait for the surgeon: "a look between him and his wife decided what was to be done" (P, p. 120). Captain Wentworth and Anne do not have Captain and Mrs. Harville's unspoken communication, and they do not have Emma and Mr. Knightley's history of strategic teamwork, but to his credit, Captain Wentworth creates that partnership on the spot by explicitly asking for help and following Anne's instructions; he is the kind of man who does not mind asking for directions.

One might say that they implicitly worked together earlier, when Captain Wentworth gets his sister Mrs. Croft to insist that Anne join her and her husband in their carriage. Anne's role is not completely passive: she must accept Mrs. Croft's request, and she knows that by stepping into the carriage, she acknowledges not just Captain Wentworth's courtesy but his tenderness. Anne is a willing accomplice in her own manipulation.

Even earlier, Anne and Captain Wentworth had a shared understanding, an "exquisite felicity," when they first fell in love when Anne was nineteen (P, p. 28). But Anne had broken it off in a way particularly damaging to their strategic partnership. Anne allowed a third party, Lady Russell, to intrude, and even worse, Anne believed that it was to his advantage that they not marry: "The belief of being prudent, and self-denying principally for *his* advantage, was her chief consolation, under the misery of a parting" (P, p. 30). Just to make herself feel better about rejecting him, Anne presumed to know his interests better than he knew them himself. Anne does not accept his direct avowal of what he truly desires: "she had to encounter all the additional pain of opinions, on his side, totally unconvinced and unbending" (P, p. 30). Eight years later, Anne gets what she deserves when, left with only her own doubts and delusions, she grasps for the flimsiest evidence of his true feelings: "She could not understand his present feelings, whether he were really suffering much from disappointment or not; and till that point were settled, she could not be quite herself" (P, p. 193). But after Anne reads Captain Wentworth's letter and they finally understand each other, the first thing they do, before even talking, is jointly put on a show for Charles Musgrove when he asks if it is alright for Captain Wentworth to escort Anne home instead: "There could be only a most proper alacrity, a most obliging compliance for public view; and smiles reined in and spirits dancing in private rapture" (P, p. 261).

Elizabeth and Mr. Darcy's strategic partnership, unlikely at first, is nevertheless detectable. While Jane recovers from her illness at Nether-field, Mrs. Bennet arrives to check up on her, and immediately displays her foolishness by taking offense with Mr. Darcy's comment that "[i]n a country neighbourhood you move in a very confined and unvarying society" (PP, p. 47). "[B]lushing for her mother," Elizabeth speaks for Mr. Darcy, saying, "He only meant that there were not such a variety of people to be met with in the country as in town, which you must acknowledge to be true" (PP, p. 47). When Mrs. Bennet snorts, "I know we dine with four and twenty families," Elizabeth tries to change the subject to Charlotte Lucas, which only results in Mrs. Bennet declaring that Jane is much prettier (PP, p. 48). Mrs. Bennet goes on to recall that a suitor once wrote love poetry to Jane, and Elizabeth tries to change the subject again by interjecting, "I wonder who first discovered the efficacy of poetry in driving away love!" Now Mr. Darcy keeps the diversion going by replying, "I have been used to consider poetry as the *food* of love." Elizabeth continues, "Of a fine, stout, healthy love it may.... But if it be only a slight, thin sort of inclination, I am convinced that one good sonnet will starve it entirely away" (PP, p. 49). This is as far as the conversation goes; Mr. Darcy can only smile in response. At least to some small degree, Mr. Darcy acts with Elizabeth like Mr. Knightley does with Emma, following her lead to try to prevent a parent from embarrassing herself or himself further. Even though Elizabeth dislikes Mr. Darcy, she is willing to defend him based on her understanding of his intentions, and Mr. Darcy reciprocates by acting, albeit briefly, on her intentions.

Later that day, Elizabeth and Mr. Darcy discuss how Mr. Bingley makes decisions. Mr. Darcy tells Mr. Bingley, "Your conduct would be quite as dependant on chance as that of any man I know; and if, as you were mounting your horse, a friend were to say, 'Bingley, you had better stay till next week,' you would probably do it, you would probably not go—and, at another word, might stay a month" (PP, p. 53). In defense of Mr. Bingley, Elizabeth says to Mr. Darcy, "To yield readily—easily—to the *persuasion* of a friend is no merit with you" (PP, p. 54). Mr. Darcy fleshes out the scenario by saying that the hypothetical friend has not offered any particular argument that Mr. Bingley should stay. Elizabeth replies that the scenario is still incompletely specified: "A regard for the requester would often make one readily yield to a request, without waiting for arguments to reason one into it.... We may as well wait, perhaps, till the circumstance occurs, before we discuss the discretion of his behaviour thereupon" (PP, p. 54). Mr. Darcy agrees that more detail is necessary: "Will it not be advisable, before we proceed on this subject, to arrange with rather more precision the degree of importance which is to appertain to this request, as well as the degree of intimacy subsisting between the parties?" (PP, p. 55). For Mr. Darcy and Elizabeth (and Austen), this is a real discussion; they take an interest in how people make decisions in a given scenario and whether those decisions are sensible. But Mr. Bingley doesn't take it seriously, joking that they must also specify the friend's height and size. Mr. Darcy smiles, "but Elizabeth thought she could perceive that he was rather offended; and therefore checked her laugh." Mr. Darcy reproaches Mr. Bingley for trying to end the discussion, but Elizabeth suggests that Mr. Darcy return to his task of writing to his sister, and "Mr. Darcy took her advice, and did finish his letter" (PP, p. 55). Again, Elizabeth tries to understand Mr. Darcy's feelings and act accordingly by not laughing, and Mr. Darcy in the end follows Elizabeth's instructions.

In these minor interactions, Elizabeth and Mr. Darcy establish a pattern of working together, despite her dislike. When Elizabeth refuses Mr. Darcy's proposal, she clearly states her reasons for doing so and thus Mr. Darcy is able to take them as instructions. Elizabeth blames Mr. Darcy for persuading Mr. Bingley not to marry Jane, but in his letter to Elizabeth, Mr. Darcy writes, "I remained convinced... that though she received his attentions with pleasure, she did not invite them by any participation of sentiment.... I must have been in an error" (PP, p. 219). Like Captain Wentworth agreeing that Captain Benwick knows Lyme better, Mr. Darcy admits that Elizabeth knows her sister better. Elizabeth knows Jane and Mr. Darcy knows Mr. Bingley, and thus Elizabeth's refusal and Mr. Darcy's letter in response form a joint plan to get Jane and Mr. Bingley back together. Their earlier joint analysis of how Mr. Bingley makes decisions turns out to be quite relevant; all Mr. Darcy has to do is assure him that Jane indeed loves him. Later, when Lydia runs off with Wickham, Mr. Darcy calls upon his hard-earned knowledge of an earlier elopement attempt of Wickham's, with his own sister, Georgiana. Mr. Darcy first bribes Mrs. Younge, Georgiana's former governess and Wickham's earlier accomplice, to disclose Wickham's location, and next bribes Wickham himself to marry Lydia. Mr. Darcy wins Elizabeth by demonstrating his fitness as a strategic partner.

Edward Ferrars is admittedly not much of a strategic partner; Marianne is correct to say that "[h]is eyes want all that spirit, that fire, which at once announce virtue and intelligence" (SS, p. 20). When Colonel Brandon offers Edward a living after he is cast off by his family, Edward thanks Elinor for her suspected role in the gesture, but that's about the extent of his strategic thinking. Most of Edward's strategic ability is occupied with preventing himself from unintentionally revealing his secret engagement with Lucy Steele. For example, when Marianne spots a ring he is wearing that contains a braid of Lucy's hair, he quickly and awkwardly agrees with Marianne's suggestion that the hair must be from his sister Fanny (Marianne thinks to herself that the hair must be Elinor's). When Marianne jokingly calls Edward Ferrars reserved, he stutters, "Reserved!—how, in what manner? What am I to tell you? What can you suppose?" (SS, p. 109).

Colonel Brandon and Elinor are the true strategic partners; the "good understanding between the Colonel and Miss [Elinor] Dashwood seemed rather to declare that the honours of the mulberry-tree, the canal, and the yew arbour, would all be made over to" Elinor instead of Marianne (SS, p. 246). Colonel Brandon asks Elinor for help with Marianne: "Does your sister make no distinction in her objections against a second attachment? . . . Are those who have been disappointed in their first choice, whether from the inconstancy of its object, or the perverseness of circumstances, to be equally indifferent during the rest of their lives?" (SS, p. 67). Colonel Brandon, who had loved once before, is looking for a loophole that might enable his suit. When Mrs. Jennings is on the verge of teasing the name of Elinor's favorite out of the youngest Dashwood sister Margaret, Lady Middleton, out of "great dislike of all such inelegant subjects of raillery," switches the subject to the weather, and "[t]he idea however started by her, was immediately pursued by Colonel Brandon, who was on every occasion mindful of the feelings of others; and much was said on the subject of rain by both of them" (SS, p. 72). Again like Mr. Knightley, who follows Emma's lead and changes the subject to prevent direct conflict between his brother and Mr. Woodhouse, Colonel Brandon jumps on the opportunity, presented by Lady Middleton, to prevent Elinor's further embarrassment. When Marianne falls ill, it is more than natural for Elinor to rely on Colonel Brandon to fetch their mother.

Once the strategic partnership between Colonel Brandon and Elinor is well rooted, all that is necessary is to graft Marianne onto it. Mrs. Dashwood has the "wish of bringing Marianne and Colonel Brandon together.... [T]o see Marianne settled at the mansion-house was equally the wish of Edward and Elinor.... With such a confederacy against her ... what could she do?" (SS, pp. 428–29).

Once released from his engagement to Lucy, Edward is able to think strategically together with Elinor, at least in retrospect. Elinor wonders how Lucy managed to marry Robert Ferrars: "by what attraction Robert could be drawn on to marry a girl ... already engaged to his brother, and on whose account that brother had been thrown off by his family—it was beyond her comprehension to make out" (SS, p. 412). Edward suggests that "Lucy perhaps at first might think only of procuring his good offices in my favour. Other designs might afterward arise" (SS, p. 413). Edward still believes that Lucy must have had some genuine affection for him while they were engaged, because "what fancied advantage it could be to her, to be fettered to a man for whom she had not the smallest regard, and who had only two thousand pounds in the world. She could not foresee that Colonel Brandon would give me a living" (SS, p. 416). Elinor replies, "[S]he lost nothing by continuing the engagement, for she has proved that it fettered neither her inclination nor her actions.... [I]f nothing more advantageous occurred, it would be better for her to marry *you* than be single" (SS, p. 416). Elinor and Edward talk in completely strategic terms about Lucy's preferences over her alternatives and her anticipation of others' actions.

This post-game recap, in which a couple reviews the choices and motivations of others and themselves, is often the moment of greatest intimacy. On their walk to the Allen residence, Henry Tilney's proposal to Catherine is intertwined with an explanation of his father's and his own actions: "Some explanation on his father's account he had to give; but his first purpose was to explain himself, and before they reached Mr. Allen's grounds he had done it so well, that Catherine did not think it could ever be repeated too often" (NA, p. 252). Mr. Knightley had gone to visit Emma with the sole intention of consoling her after the news of Frank Churchill's engagement with Jane Fairfax. Emma explains that "my vanity was flattered, and I allowed his attentions.... [But] I have never been attached to him. And now I can tolerably comprehend his behaviour. He never wished to attach me. It was merely a blind to conceal his real situation with another" (E, pp. 465–66). This discussion of Frank Churchill's strategic actions naturally involves a discussion of Emma's own preferences, and "[t]he delightful assurance of her total indifference towards Frank Churchill ... had given birth to the hope" which moves Mr. Knightley toward confessing his own feelings

(E, p. 471). Fanny and Edmund Bertram had talked about Mary Crawford's faults many times, but their closest moment is when Edmund tells Fanny about his last meeting with Mary. Mary had suggested that after her brother Henry had run off with Edmund's sister, the married Maria, the Bertrams should not raise a fuss: "What I advise is, that your father be quiet.... If by any officious exertions of his, she is induced to leave Henry's protection, there will be much less chance of his marrying her" (MP, pp. 528–29). Edmund cannot believe that she is "recommending to us a compliance, a compromise, an acquiescence, in the continuance of the sin," and Fanny can only add that Mary seemed more willing to accept Edmund after hearing of the seriousness of Tom Bertam's illness, which would make Edmund heir if Tom died (MP, p. 529). Knowledge of Mary Crawford's motives is especially hurtful to Edmund, but discussing them with Fanny propels Edmund and Fanny toward their own union.

When Elizabeth Bennet thanks Mr. Darcy for quietly underwriting Lydia's marriage, adding that if her family knew, they would all be equally grateful, Mr. Darcy explains, "[Y]our *family* owe me nothing. Much as I respect them, I believe I thought only of *you*. ... My affections and wishes are unchanged, but one word from you will silence me on this subject for ever" (PP, p. 406). Thus Mr. Darcy's declaration of love follows from an explanation of his motivations for acting strategically. This time Elizabeth accepts, and together they immediately go through a play-by-play analysis including details such as what Mr. Darcy thought when Elizabeth unexpectedly showed up with the Gardiners at Pemberley, Elizabeth's surprise at being treated with such warmness there, Mr. Darcy's motivation "to shew you, by every civility in my power, that I was not so mean as to resent the past; and I hoped to obtain your forgiveness, to lessen your ill opinion, by letting you see that your reproofs had been attended to," and finally how Mr. Darcy personally verified, by direct observation, that Jane had real affection for Mr. Bingley, and persuaded Mr. Bingley of this fact (PP, p. 410). Later, Elizabeth cannot help engaging in still more post-game discussion. "[R]ising to playfulness again, she wanted Mr. Darcy to account for his having ever fallen in love with her" (PP, p. 421). Elizabeth asks, "[D]id you admire me for my impertinence? ... Why, especially, when you called, did you look as if you did not care about me? ... I wonder when you *would* have spoken, if I had not asked you!" (PP, pp. 421–22). After Anne and Captain Wentworth realize their affection for each other, "they could indulge in those retrospections and acknowledgements, and especially in those explanations of what had directly preceded the present moment, which were so poignant and so ceaseless in interest" (P, p. 262).

A perfect understanding does not diminish the sweetness of replaying how it was achieved.

Austen's dubious as well as heroic couples are cemented through strategic partnership. Maria Bertram and Henry Crawford jointly maneuver to get the right parts in *Lovers' Vows* so that they can frequently rehearse together. The widow Mrs. Clay had been angling to marry Anne Elliot's father, Sir Walter, and Mr. Elliot, Sir Walter's heir presumptive, had been trying to secure his inheritance by preventing Sir Walter from producing a son; he only wanted to marry Anne in order to better monitor Sir Walter. Mrs. Clay and Mr. Elliot are brought together by their competing manipulation of Sir Walter, and it is not surprising that she would be "next heard of as established under his protection in London" (P, p. 273).

Strategic partnership is also essential to friendships between women. Mrs. Smith's closeness to Anne is demonstrated by how she finds a way, even as an invalid confined to her room, to establish informants: Mr. Elliot tells everything to Colonel Wallis, whose wife shares a nurse with Mrs. Smith, and thereby Mrs. Smith is able to warn Anne of Mr. Elliot's true intentions. In contrast, Anne's own sisters are poor strategists and thus Anne cannot be truly intimate with them, even though she wishes she could. When the Musgrove sisters come by and ask if Anne and her sister Mary would like to join them for a walk, Mary agrees even though "Anne felt persuaded, by the looks of the two girls, that it was precisely what they did not wish.... [She] thought it best to accept the Miss Musgroves' much more cordial invitation to herself to go likewise, as she might be useful in turning back with her sister, and lessening the interference in any plan of their own" (P, p. 89). Mary is bad at reading facial expressions and social situations. When Louisa Musgrove falls at Lyme, Mary's uselessness is clearly demonstrated: " 'She is dead! she is dead!' screamed Mary, catching hold of her husband" (P, p. 118). After Captain Wentworth proposes that Anne should stay and take care of Louisa, Mary messes everything up by insisting that as sister-in-law she should stay with Louisa instead; "Anne had never submitted more reluctantly to the jealous and ill-judging claims of Mary" (P, p. 124). Anne's other sister, Elizabeth, is even worse; when Anne warns her that Mrs. Clay might be trying to win their father's affections and should not accompany their relocation to Bath, "Elizabeth could not conceive how such an absurd suspicion should occur to her" (P, p. 37).

Since strategic partnership is so important to women's friendship, the severest reproach is to accuse a friend of being reserved, of not sharing vital information. When Elizabeth Bennet tells Jane of her engagement with Mr. Darcy, Jane replies, "[Y]ou have been very sly, very reserved with me. How little did you tell me of what passed at Pemberley and Lambton!" (PP, p. 415). When Elinor and Marianne arrive in London, Elinor, suspecting that Marianne is trying to contact Willoughby, asks Marianne if she is expecting a letter. When Marianne replies unspecifically, Elinor says, "[Y]ou have no confidence in me, Marianne." Marianne answers, "Nay, Elinor, this reproach from *you*— you who have confidence in no one!" (SS, p. 193). Emma hates Jane Fairfax because "[t]here was no getting at her real opinion. Wrapt up in a cloak of politeness, she seemed determined to hazard nothing. She was disgustingly, was suspiciously reserved" (E, p. 180).

Among men, Captain Harville is a skilled wingman for Captain Wentworth, crucially drawing out Anne's declaration of constancy in his friend's presence. Even the manipulable Mr. Bingley can act strategically for Mr. Darcy's benefit. When he and Mr. Darcy visit the Bennets, he asks Mrs. Bennet, "[H]ave you no more lanes hereabouts in which Lizzy may lose her way again to-day," and thus Mrs. Bennet, to get Mr. Darcy out of Mr. Bingley's way, suggests that Mr. Darcy, Elizabeth, and Kitty take a walk together. Mr. Bingley only has to add, "It may do very well for the others... but I am sure it will be too much for Kitty," freeing Elizabeth and Mr. Darcy to plan how they will ask Elizabeth's parents for their consent (PP, p. 416).

Between brother and sister, William Price tells Fanny that he worries about his promotion to lieutenant, and Fanny assures him that Sir Thomas "says nothing, but... will do everything in his power to get you made" (MP, p. 290). Later when Fanny is suffering under Henry Crawford's proposal, William, "knowing her wish on the subject... would not distress her by the slightest allusion" (MP, p. 434). In contrast, the gap between the Dashwood sisters and their half-brother John Dashwood is evident by the crudeness of his machinations, as when he goads Elinor: "[I]t would give me great pleasure to call Colonel Brandon brother. His property here, his place, his house, every thing is in such respectable and excellent condition!—and his woods!—I have not seen such timber any where in Dorsetshire" (SS, p. 425).

Between father and daughter, when Mary Bennet sings dreadfully at the Netherfield ball, Elizabeth "looked at her father to entreat his interference, lest Mary should be singing all night. He took the hint" (PP, pp. 112–13). When Mrs. Bennet tells Elizabeth that she will never speak to her again if she does not accept Mr. Collins's proposal, Mr. Bennet gives his own answer in the form of a game, which as a fellow strategist he knows Elizabeth will understand: "From this day you must be a stranger to one of your parents.—Your mother will never see you again if you do *not* marry Mr. Collins, and I will never see you again if you *do*" (PP, p. 125).

Strategic partners jointly manipulate horses as well as people. Mary Crawford asks Edmund for riding lessons, and Fanny rightly observes them with trepidation: "Edmund was close to her, he was speaking to her, he was evidently directing her management of the bridle, he had hold of her hand" (MP, p. 79). Riding in John Thorpe's carriage, Catherine spots the Tilneys and pleads to stop. Showing no commitment to the team concept and dooming any chance he may have had with Catherine, "Mr. Thorpe only laughed, smacked his whip, encouraged his horse, made odd noises, and drove on" (NA, p. 86). In comparison, Admiral and Mrs. Croft, Austen's "prototype for... successful marriages" (Mellor 1993, p. 57), drive their carriage as a team and "function as a cognitive unit" (Palmer 2010, p. 152). When Mrs. Croft spots an obstacle, "by coolly giving the reins a better direction herself, they happily passed the danger; and by once afterwards judiciously putting out her hand, they neither fell into a rut, nor ran foul of a dung-cart; and Anne, with some amusement at their style of driving... imagined [this] no bad representation of the general guidance of their affairs" (P, p. 99).

STRATEGIZING ABOUT YOURSELF

If game theory can be criticized for being too atomistic, it can also be criticized for not being atomistic enough: an individual is often usefully understood as a confederation of different parts or "selves" (see, for example, Ainslie 1992, Benhabib and Bisin 2005, Fudenberg and Levine 2006, O'Donoghue and Rabin 2001). A person can understand and manipulate herself just as she can understand and manipulate others. For example, while Anne Elliot waits inside with her sister and Mrs. Clay for Lady Dalrymple's carriage because it is raining, she is startled to see Captain Wentworth walking down the street: "Her start was perceptible only to herself; but she instantly felt that she was the greatest simpleton in the world, the most unaccountable and absurd! For a few minutes she saw nothing before her. It was all confusion. She was lost; and when she had scolded back her senses, she found the others still waiting for the carriage" (P, p. 190). Here Anne has a "cognitive" or "executive" self and an "embodied" or "sensing" simpleton self. Anne's executive self observes the reflex start of her embodied self, makes sure that no one else sees her embodied self's reaction but is still embarrassed about it, and scolds her embodied, sensing self into functioning again. "She now felt a great inclination to go to the outer door; she wanted to see if it rained. Why was she to suspect herself of another motive? Captain Wentworth must be out of sight. She left her seat, she would go, one half of her should not be always so much wiser than the other half, or always suspecting the other of being worse than it was. She would see if it rained" (P, p. 190). Now Anne is explicitly divided into two halves; according to Richardson (2002, p. 149), Anne is "split between a superintending conscious self and a potentially unruly, desiring, unconscious other." The "feeling" Anne has a great inclination to go outside, has a perfectly good excuse, and rebels against a suspicious and purportedly wiser "executive" Anne who cares about appearances. Even though Austen generally values self-command, in this example the commanded self successfully negotiates with the commanding self.

Indeed, Austen emphasizes that the relationship between a person's selves is somewhat more complicated than a hierarchical chain of command. When Admiral and Mrs. Croft first visit Kellynch Hall to see whether they would like to rent it, "Anne found it most natural to take her almost daily walk to Lady Russell's, and keep out of the way till all was over; when she found it most natural to be sorry that she had missed the opportunity of seeing them" (P, p. 34). Here the "natural" part of Anne, which takes a walk out of habit, offers an excuse and apology that the "executive" part of Anne accepts. Here Anne's executive self does not require Anne's habitual self to resolutely make a conscious choice whether to see or avoid the Crofts; Anne's executive self lets this one slide. When the Crofts move in and Captain Wentworth is soon to visit, Anne's executive self resumes its usual role of commanding Anne's sensing self: the Musgrove family's talking about Captain Wentworth "was a new sort of trial to Anne's nerves.... to which she must enure herself.... [S]he must teach herself to be insensible on such points" (P, p. 56). After Anne and Captain Wentworth finally meet, Anne's executive self comes down hard on Anne's curious sensing self ("Now, how were his sentiments to be read?... And the next moment she was hating herself for the folly which asked the question") but cannot ultimately restrain it: "perhaps her utmost wisdom might not have prevented" her wondering whether he still finds her attractive (P, p. 65).

Anne is particularly anxious that Lady Russell not see her and Captain Wentworth together: "were Lady Russell to see them together, she might think that he had too much self-possession, and she too little" (P, p. 100). Self-possession is largely an internalization of how one is observed by others, especially superiors. After Louisa Musgrove's fall at Lyme, Anne must tell Lady Russell what happened, and "Captain Wentworth's name must be mentioned.... She could not speak the name, and look straight forward to Lady Russell's eye, till she had adopted the expedient of telling her briefly what she thought of the attachment between him and Louisa. When this was told, his name distressed her no longer" (P, p. 135). Here the cognitive, executive Anne does not reprimand or command but rather assists the embarrassed, feeling Anne by devising a specific strategy to keep Lady Russell from suspecting that she still cares for Captain Wentworth.

After Anne and Captain Wentworth finally understand their feelings for each other, Anne "re-entered the house so happy as to be obliged to find an alloy in some momentary apprehensions of its being impossible to last. An interval of meditation, serious and grateful, was the best corrective of every thing dangerous in such high-wrought felicity; and she went to her room, and grew steadfast and fearless in the thankfulness of her enjoyment" (P, p. 266). In this moment of maximum happiness, perhaps the feeling Anne needs the executive Anne to tell her to slow down and assure her that the happiness will persist. Perhaps Anne, even after being victimized for eight years by her executive self's excessive caution, cannot throw her executive self off completely and cannot accept extreme happiness without somehow regulating it. Perhaps the feeling Anne, overflowing with victory, magnanimously finds a harmless task for the executive Anne: the creation of a cover story for her elation so that it can be understood by others as the more socially acceptable gratitude.

Austen interestingly compares Elinor and Marianne Dashwood's different self-management strategies. Elinor knows how to govern her feelings but this "was a knowledge which her mother had yet to learn, and which one of her sisters had resolved never to be taught" (SS, p. 7). Marianne's not restraining her emotions results from her conscious choice not to learn how, not from socialization or natural incapacity. When Marianne and Willoughby meet and fawn over each other, "Elinor could not be surprised at their attachment. She only wished that it were less openly shewn; and once or twice did venture to suggest the propriety of some self-command to Marianne" (SS, p. 63). For Elinor, self-command is largely about controlling how others see you. This is true for Marianne also, but instead of squelching her feeling self, Marianne amplifies it for public consumption. After Willoughby suddenly departs, she broadcasts a "violent sorrow which Marianne was in all probability not merely giving way to as a relief, but feeding and encouraging as a duty" (SS, p. 90). Marianne controls herself, but turns the volume up, not down. Indeed, when Edward Ferrars in turn leaves for no apparent reason, Elinor subdues her feelings and does "not adopt the method so judiciously employed by Marianne, on a similar occasion, to augment and fix her sorrow.... Their means were as different as their objects, and equally suited to the advancement of each" (SS, p. 120).

In other words, Austen argues that self-management strategies are a matter of choice, not temperament; Elinor's and Marianne's strategies differ because they have different goals. Elinor's goal is to keep her mother and sister from worrying about her, while Marianne's is to convince the largest possible audience of the depth of her own love, best proven by the uncontrollability of her suffering. Indeed, to Marianne, Elinor's subduing of her feelings only indicates their shallowness: "[W]ith strong affections it was impossible, with calm ones it could have no merit" (SS, p. 121). After Marianne receives the letter from Willoughby saying that he will marry Miss Grey, Marianne's "torrent of unresisted grief" is too much even for sisterly Elinor, who commands: "Exert yourself, dear Marianne...if you would not kill yourself and all who love you. Think of your mother; think of her misery while *you* suffer; for her sake you must exert yourself." But Marianne thinks Elinor doubts her seriousness: "Oh! how easy for those, who have no sorrow of their own to talk of exertion! Happy, happy Elinor, *you* cannot have an idea of what I suffer" (SS, p. 211). Since appealing to their mother doesn't work, Elinor next appeals to public appearances: "Whoever may have been so detestably your enemy, let them be cheated of their malignant triumph, my dear sister, by seeing how nobly the consciousness of your own innocence and good intentions supports your spirits." But Marianne wants the world to know: "I care not who knows that I am wretched. The triumph of seeing me so may be open to all the world" (SS, p. 215). Marianne takes the extreme step of allowing herself to fall ill and almost die, and only afterward vows to moderate herself. But as mentioned in chapter 5, Marianne's amplification strategy, once set to maximum drama, basically works: it does not give her Willoughby but it does make him show up in the middle of the night to confess his true feelings, and for that matter allows Colonel Brandon to bring Mrs. Dashwood and along the way receive a mother's encouragement to pursue Marianne himself.

Elinor's and Marianne's strategies need not conflict. After the Dashwoods think that Edward Ferrars has married Lucy Steele, Edward comes to visit. As he arrives on horseback, Elinor, girding her loins, says to herself, "I *will* be calm; I *will* be mistress of myself." Again concerned with appearances, Elinor wishes Marianne and her mother to "understand that she hoped no coolness, no slight, would appear in their behaviour to him" (SS, p. 406). But when Edward says that Lucy is not Mrs. Edward Ferrars but Mrs. Robert Ferrars, Elinor "could sit it no longer. She almost ran out of the room, and as soon as the door was closed, burst into tears of joy, which at first she thought would never cease. Edward, who had till then looked any where, rather than at her, saw her hurry away, and perhaps saw—or even heard, her emotion; for immediately afterwards he fell into a reverie" (SS, p. 408). As Marianne would predict, Elinor's lack of control, her running out of the room, is what reveals to Edward her true feelings and sends him aloft. To be fair, Elinor does manage to make it to another room and close the door before bursting into tears; perhaps the heroic exertion evident in this attempt at self-control is what best convinces Edward of her affection. Thus Marianne's and Elinor's self-management strategies are both validated in combination.

One reason to strategize about yourself, Austen notes, is if you anticipate that one of your selves might be biased. As Elizabeth tells Mr. Darcy, "It is particularly incumbent on those who never change their opinion, to be secure of judging properly at first" (PP, p. 105); the present self must judge carefully, anticipating the unreasonably stubborn future self. One common kind of error is "confirmation bias," a bias toward arguments consistent with what you already believe. For example, despite his doubts, Sir Thomas believes that his daughter Maria's marriage to Mr. Rushworth will be successful: "Sir Thomas was satisfied; too glad to be satisfied perhaps to urge the matter quite so far as his judgment might have dictated to others. . . . [He was] very happy to think any thing of his daughter's disposition that was most favourable for the purpose" (MP, p. 235). When Anne argues with Captain Harville over which sex is more constant, Anne suggests that they might never agree because "[w]e each begin probably with a little bias towards our own sex, and upon that bias build every circumstance in favour of it which has occurred within our own circle" (P, p. 255). In a meta-analysis of ninety-one experimental studies, "People are almost two times . . . more likely to select information congenial rather than uncongenial to their pre-existing attitudes, beliefs, and behaviors" (Hart, Albarracín, Eagly, Brechan, Lindberg, and Merrill 2009, p. 579; see also Baron 2008, p. 215).

Austen's strategically thoughtful people try to be self-critically aware of potential bias. In his letter to Elizabeth, Mr. Darcy explains that he truly believed that her sister Jane did not have any real interest in Mr. Bingley, defending himself against confirmation bias by acknowledging its possibility: "That I was desirous of believing her indifferent is certain,—but I will venture to say that my investigations and decisions are not usually influenced by my hopes or fears.—I did not believe her to be indifferent because I wished it" (PP, p. 220). Mr. Knightley suspects Frank Churchill, and cannot "avoid observations which, unless it were like Cowper and his fire at twilight, 'Myself creating what I saw,' brought him yet stronger suspicion of there being a something of private liking, of private understanding even, between Frank Churchill and Jane" (E, p. 373). Colonel Brandon states that "where the mind is perhaps rather unwilling to be convinced, it will always find something to support its doubts" (SS, p. 197).

Edmund Bertram freely admits that his choice of the clergy as a profession was biased by the fact that his father had already arranged a living for him as a clergyman: "[K]nowing that there was such a provision for me, probably did bias me. Nor can I think it wrong that it should. . . . I see no reason why a man should make a worse clergyman for knowing that he will have a competence early in life. . . . I have no doubt that I was biassed, but I think it was blamelessly" (MP, p. 127). Fanny despairs at the likelihood of Edmund's marriage to Mary Crawford, and believes that her ill opinion of Mary is fair despite her own personal interest: "there were bad feelings still remaining which made the prospect of it most sorrowful to her, independently—she believed independently of self. In their very last conversation, Miss Crawford . . . [had] still shewn a mind led astray and bewildered, and without any suspicion of being so" (MP, p. 423). Among Mary's liabilities is a lack of awareness of her own biases.

PREFERENCE CHANGE

For Austen, a change in someone's preferences is always deserving of notice and explanation. Austen is generally skeptical of preference change, seeing it as an amusing human failing to which mainly the unserious and immature are susceptible. She explores several mechanisms for preference change, including the noble mechanism of gratitude, the understandable mechanisms of being near death or in love, the slightly dubious mechanism of "reference dependence," the condemnable mechanisms of flattery and persuasion, and the absurd mechanism of self-rationalization. Also, sometimes one's preference for an alternative changes because the alternative itself significantly changes, by gaining a new feature or taking on a new social connotation.

For Austen, the most laudable mechanism for preference change is gratitude. Elizabeth's changing feelings for Mr. Darcy are chronicled through increasing gratitude. When Mr. Darcy first proposes, Elizabeth replies, "It is natural that obligation should be felt, and if I could *feel* gratitude, I would now thank you. But I cannot—I have never desired your good opinion" (PP, p. 212). After Elizabeth reads Mr. Darcy's letter answering her objections, "[h]is attachment excited gratitude, his general character respect; but she could not approve him; nor could she for a moment repent her refusal" (PP, p. 236). At first she does not feel gratitude at all; then she does, but not enough to make her change her mind. When Elizabeth visits Pemberley with the Gardiners, the housekeeper Mrs. Reynolds praises Mr. Darcy's character, and "she thought of his regard with a deeper sentiment of gratitude than it had ever raised before" (PP, p. 277). When Mr. Darcy shows up a day early and treats them with great hospitality, even introducing his sister Georgiana, Elizabeth finds that "there was a motive within her of good will which could not be overlooked. It was gratitude.—Gratitude, not merely for having once loved her, but for loving her still well enough, to forgive all the petulance and acrimony of her manner in rejecting him" (PP, p. 293). Indeed, "[i]f gratitude and esteem are good foundations of affection, Elizabeth's change of sentiment will be neither improbable nor faulty" (PP, p. 308). After Elizabeth learns that Mr. Darcy secretly underwrote Wickham's marriage to Lydia, Elizabeth expresses her gratitude ("Ever since I have known it, I have been most anxious to acknowledge to you how gratefully I feel it") and thereby initiates a fruitful conversation: Mr. Darcy expresses the constancy of his affection, and Elizabeth replies "that her sentiments had undergone so material a change . . . as to make her receive with gratitude and pleasure, his present assurances" (PP, pp. 405–406). In the final sentence of *Pride and Prejudice*, Elizabeth and Mr. Darcy "were both ever sensible of the warmest gratitude" toward the Gardiners for bringing them together at Pemberley (PP, p. 431).

Gratitude is powerful, but maybe it operates simply. Instead of gratitude creating affection, perhaps a person is simply more likely to love you if she thinks that you love her. In other words, it's not that one person changes the preferences of another but rather that two people are in a coordination problem, like Beatrice and Benedick in chapter 2. Henry Tilney falls in love with Catherine Morland, but Austen "must confess that his affection orginated in nothing better than gratitude, or, in other words, that a persuasion of her partiality for him had been the only cause of giving her a serious thought" (NA, pp. 252–53). Here Austen equates gratitude with simply knowing that someone else likes you. Emma, thinking of Frank Churchill, "could not doubt his having a decidedly warm admiration, a conscious preference of herself; and this persuasion, joined to all the rest, made her think that she *must* be a little in love with him" (E, pp. 282–83). Emma does not feel grateful to Frank Churchill, but believing that he loves her is enough to make her wonder whether she loves him. Mellor (1993, p. 56) argues that male gratitude, specifically Henry Tilney's toward Catherine, is based on vanity, but vanity seems to better explain Emma's reaction; Mellor also argues that gratitude is something that inferiors, typically female, feel toward superiors, but this does not explain Mr. Darcy's gratitude toward the Gardiners.

Feelings of gratitude recur in *Pride and Prejudice*, but so do coordination problems. Charlotte Lucas, talking to Elizabeth about her sister Jane, agrees that love is a coordination problem, in which your target likes you more the more he knows that you like him: "[T]here are very few of us who have heart enough to be really in love without encouragement.... Bingley likes your sister undoubtedly; but he may never do more than like her, if she does not help him on" (PP, p. 24). Indeed, what stops Mr. Bingley's suit is Mr. Darcy convincing him that Jane is indifferent; once Mr. Darcy tells Mr. Bingley that he had been mistaken and now sincerely believes that Jane loves him, this is enough to reattach him to Jane. Like Beatrice toward Benedick, Elizabeth behaves impertinently toward Mr. Darcy in the first place because she expects him to be impertinent toward her; as she tells Charlotte, "[H]e has a very satirical eye, and if I do not begin by being impertinent myself, I shall soon grow afraid of him" (PP, p. 26; see also Knox-Shaw 2004, p. 88 on Elizabeth as "Beatrice-like"). After their visit to Pemberley, Elizabeth and Mrs. Gardiner long to talk to each other about Mr. Darcy, but each wants the other to go first: "Elizabeth was longing to know what Mrs. Gardiner thought of him, and Mrs. Gardiner would have been highly gratified by her niece's beginning the subject" (PP, p. 300). According to Doody (1988, p. 231), the entire point of Frances Burney's novel *Camilla* (which Austen had read, and even mentions in *Northanger Abbey*) is to express frustration with a courtship system in which everyone is instructed to "gain the other's heart before giving anything of yours."

A near-death experience can quite effectively change a person's preferences. After Louisa Musgrove recovers from her fall, she turns into "a person of literary taste, and sentimental reflection.... The day at Lyme, the fall from the Cobb, might influence her health, her nerves, her courage, her character to the end of her life" (P, p. 182). Tom Bertram recovers from illness and "became what he ought to be, useful to his father, steady and quiet, and not living merely for himself" (MP, p. 534). After Marianne Dashwood recovers, she declares that she will "divide every moment between music and reading. I have formed my plan, and am determined to enter on a course of serious study" (SS, p. 388). Near-death experiences work well but are difficult to apply except accidentally.

A similar mechanism for altering one's preferences is being in love, another intense emotional state. Mr. Knightley visits Emma thinking that she must be crushed by the revelation of Frank Churchill's secret engagement to Jane Fairfax, but Emma makes clear that she had never been attached to Frank Churchill. Mr. Knightley concludes logically that Frank Churchill is therefore not entirely dastardly, but what truly changes his opinion is his elation once Emma accepts his own affection: "Within half an hour, he had passed from a thoroughly distressed state of mind, to something so like perfect happiness, that it could bear no other name.... She was his own Emma, by hand and word, when they returned into the house; and if he could have thought of Frank Churchill then, he might have deemed him a very good sort of fellow" (E, pp. 471–72). Anne Elliot is aware of this mechanism. Still hurt by Anne's rejection eight years earlier, Captain Wentworth, when first seeing Anne again, tells Henrietta that Anne is altered beyond recognition, and Anne hears of his opinion. But after they finally realize their feelings for each other, Captain Wentworth tells Anne that "to my eye you could never alter," and "the value of such homage was inexpressibly increased to Anne, by comparing it with former words, and feeling it to be the result, not the cause of a revival of his warm attachment" (P, p. 264). The height of Wentworth's love is demonstrated by how strongly it reverses his judgment. Captain Wentworth is sincere but slightly silly (like Mr. Knightley, perhaps endearingly so); steadfast Anne's preferences would never similarly reverse.

Another mechanism for preference change is "reference dependence": the payoff of an outcome can depend on the reference point of comparison or the status quo to which one is accustomed (Tversky and Kahneman 1991). For example, Fanny's brother William, now a lieutenant, visits and is excited to show Fanny his new uniform, but cannot because he is not on duty. Edmund Bertram "conjectured that before Fanny had any chance of seeing it, all its own freshness, and all the freshness of its wearer's feelings, must be worn away.... [F]or what can be more unbecoming, or more worthless, than the uniform of a lieutenant, who has been a lieutenant a year or two, and sees others made commanders before him?" (MP, pp. 424–25). A lieutenant's uniform is a real achievement for a midshipman, but once one is accustomed to being a lieutenant, it becomes "a badge of disgrace" (MP, p. 424).

Austen includes six more examples of reference dependence in *Mansfield Park* in the chapters after Fanny refuses Henry Crawford's proposal; to some degree, reference dependence drives the plot. The irksome Mary Crawford keeps trying to persuade Fanny to accept, but is unexpectedly kind when saying farewell, and Fanny's "disposition was peculiarly calculated to value a fond treatment... from having hitherto known so little of it" (MP, p. 421). After the Crawfords leave, Sir Thomas hopes that Fanny will value Henry more, now that he is not around: "he entertained great hope that his niece would find a blank in the loss of those attentions which at the time she had felt, or fancied an evil" (MP, p. 422). Since Fanny does not change her mind about Henry Crawford, Sir Thomas makes Fanny visit her former home in Portsmouth, hoping that "a little abstinence from the elegancies and luxuries of Mansfield Park would... incline her to a juster estimate of the value of that home of greater permanence, and equal comfort, of which she had the offer" (MP, p. 425). Sir Thomas expects that Henry's proposal will look better from the status quo of Portsmouth than from the status quo of Mansfield Park. Indeed, when Fanny sits in her noisy crowded home at Portsmouth, "[t]he elegance, propriety, regularity, harmony—and perhaps, above all, the peace and tranquillity of Mansfield, were brought to her remembrance every hour of the day, by the prevalence of every thing opposite to them *here*" (MP, p. 453). Only a few weeks earlier at Mansfield, Fanny had wished Mary Crawford to go away, but when Fanny receives a letter from her, "Here was another strange revolution of mind!—She was really glad to receive the letter when it did come. In her present exile from good society . . . a letter from one belonging to the set where her heart lived, written with affection, and some degree of elegance, was thoroughly acceptable" (MP, p. 455). Finally, after relations between the two families are ruptured when Henry Crawford runs off with the married Maria, Mary Crawford has difficulty finding a man "who could satisfy the better taste she had acquired at Mansfield, whose character and manners could . . . put Edmund Bertram sufficiently out of her head" (MP, p. 543). Once Mary is used to Edmund Bertram, her preference for other men is diminished.

Similarly, past failure can make a current success more desirable. After missing the Tilneys the day before because of her ride with the Thorpes and her brother, Catherine Morland sees Eleanor Tilney and goes up to talk to her, "with a firmer determination to be acquainted, than she might have had courage to command, had she not been urged by the disappointment of the day before" (NA, p. 69).

Another related mechanism is that the payoff of an alternative can depend on what it is being compared with. After Maria marries and Julia goes off to live with her, Fanny is the "only young woman in the drawing-room" (MP, p. 239) and Henry Crawford exclaims that "she is now absolutely pretty. . . . [H]er air, her manner, her tout ensemble is so indescribably improved! She must be grown two inches, at least." But Mary Crawford replies, "This is only because there were no tall women to compare her with. . . . [S]he was the only girl in company for you to notice, and you must have a somebody" (MP, p. 268). Later when Henry visits her in Portsmouth, Fanny, "[n]ot considering in how different a circle she had been just seeing him, nor how much might be owing to contrast," finds Henry more gentle and thoughtful (MP, p. 479).

Fanny, who is only human and quite young, is vulnerable to reference dependence, but this is considered a weakness, not a strength. Catherine is also quite young and reference dependence is part of her naivety, which admittedly ends up working in her favor. Henry Crawford's upgrading of Fanny because no other young women are around is just silly.

For Austen, the most condemnable mechanisms for changing a person's preferences are flattery and persuasion, which work only on the very young (and fools like Mr. Rushworth). When her brother James asks Catherine Morland if she likes John Thorpe, "instead of answering, as she probably would have done, had there been no friendship and no flattery in the case, 'I do not like him at all;' she directly replied, 'I like him very much; he seems very agreeable'" (NA, pp. 44–45). Catherine is young, and "[h]ad she been older or vainer, such attacks might have done little; but, where youth and diffidence are united, it requires uncommon steadiness of reason to resist the attraction of being called the most charming girl in the world" (NA, p. 44). Young Fanny is not invulnerable to flattery: Austen writes that "although there doubtless are such unconquerable young ladies of eighteen... as are never to be persuaded into love against their judgment by all that talent, manner, attention, and flattery can do, I have no inclination to believe Fanny one of them" (MP, pp. 269–70). Lady Russell had persuaded Anne Elliot not to marry Captain Wentworth when Anne was nineteen: "Young and gentle as she was, it might yet have been possible to withstand her father's ill-will... but Lady Russell, whom she had always loved and relied on, could not, with such steadiness of opinion, and such tenderness of manner, be continually advising her in vain" (P, pp. 29–30). It is good that flattery and persuasion work only on the young, as they are almost never beneficial.

Sometimes what looks like a preference change is not really a change at all. For example, when Mr. Collins first plans to propose to Jane Bennet, then proposes to Elizabeth, and finally proposes to Charlotte Lucas, this is not a preference change; his actions are consistent with liking Jane best, Elizabeth second, and Charlotte third.

Also, one's preference for something can change because it becomes bundled with some new feature, not because one's preference for the thing itself changes. For example, Mrs. Bennet goes from hating Mr. Darcy to loving him once he comes with the additional feature of being her son-in-law, and is not at all bothered by this seeming inconsistency: "Oh, my dear Lizzy! pray apologise for my having disliked him so much before. I hope he will overlook it. Dear, dear Lizzy. A house in town! Every thing that is charming! Three daughters married! Ten thousand a year!" (PP, pp. 419–20). When Mrs. Weston suggests that Mr. Knightley likes Jane Fairfax, Emma insists that he must never marry because then her nephew Henry, who is also Mr. Knightley's nephew, would no longer be heir. But after she accepts Mr. Knightley's proposal, Emma "was never struck with any sense of injury to her nephew Henry.... Think she must of the possible difference to the poor little boy; and yet she only gave herself a saucy conscious smile about it" (E, p. 490). If she herself is his wife and the line of inheritance will go through her own children, Emma is willing to put up with Mr. Knightley marrying.

The most interesting kind of bundling is when one's preference for something changes because it takes on a new social connotation. When Mr. Henry Dashwood dies, his wife, Mrs. Dashwood, and daughters Elinor, Marianne, and Margaret remain at Norland only as guests, because John Dashwood, his son from a previous marriage, takes ownership. They plan to move out eventually but want to stay close to their beloved home. However, when John Dashwood's wife, Fanny, insinuates to Mrs. Dashwood that her brother Edward Ferrars has higher expectations for his marriage and that Elinor must not try to "*draw him in,*" a furious Mrs. Dashwood decides to leave immediately (SS, p. 26). When the invitation to stay at Barton Cottage arrives, "[t]he situation of Barton, in a county so far distant from Sussex as Devonshire, which, but a few hours before, would have been a sufficient objection to outweigh every possible advantage belonging to the place, was now its first recommendation. To quit the neighbourhood of Norland was no longer an evil; it was an object of desire; it was a blessing, in comparison of the misery of continuing [as] her daughter-in-law's guest" (SS, p. 27). Similarly, Maria and Mr. Rushworth plan to marry when her father, Sir Thomas, returns from abroad, but in the meantime Maria shamelessly accepts the attentions of Henry Crawford. When Sir Thomas returns, Henry must act quickly if serious, but instead leaves saying nothing. Maria, who had been putting off the wedding as long as possible, now desires it immediately: "Henry Crawford had destroyed her happiness, but he should not know that he had done it; he should not destroy her credit, her appearance, her prosperity too. He should not have to think of her as pining in the retirement of Mansfield for *him,* rejecting Sotherton and London, independence and splendour for *his* sake" (MP, pp. 235–36). Maria's payoff from marrying Mr. Rushworth increases because she can thereby spite Henry Crawford.

Similarly, if one's value of something changes after learning more about it, this is not really a preference change. When Henry Crawford observes the strength of Fanny Price's affection for her brother William after he returns from years at sea, "Fanny's attractions increased—increased two-fold—for the sensibility which beautified her complexion and illumined her countenance, was an attraction in itself. He was no longer in doubt of the capabilities of her heart" (MP, p. 274). Henry is more attracted to Fanny after learning what her heart can do. After everyone knows that Willoughby rejected Marianne to marry Miss Grey, Elinor grows weary of "officious condolence" and "clamorous kindness" and appreciates Lady Middleton's self-centeredness: "It was a great comfort...to know that there was *one* who would meet her without feeling any curiosity after particulars, or any anxiety for her sister's health. Every qualification is raised at times, by the circumstances of the moment, to more than its real value" (SS, p. 245). Elinor's preference for Lady Middleton's company does not change (on later occasions, she finds her as insipid as ever); rather, Elinor discovers an unexpected contingency in which Lady Middleton's unconcern pays off.

Finally, sometimes when a person says that his own or another's preferences have changed, this is not a real change but simply a rationalization of his own behavior. After Henry Crawford, who has been flirting with both Maria and Julia Bertram, decisively indicates his preference for Maria to play the part of Agatha in *Lovers' Vows*, he first tries to assuage an enraged Julia with "gallantry and compliment," but "becoming soon too busy with his play to have time for more than one flirtation, he grew indifferent to the quarrel, or rather thought it a lucky occurrence, as quietly putting an end to what might ere long have raised expectations" (MP, p. 188). Henry initially tries to remedy the break with Julia, but out of laziness changes his mind and regards it as a good thing. Henry sees other people's preferences as freely changeable to his convenience; for example, when Mary Crawford tells him that Maria, now Mrs. Rushworth, will be mad when she hears about his proposal to Fanny, Henry replies, "Mrs. Rushworth will be very angry. It will be a bitter pill to her; that is, like other bitter pills, it will have two moments' ill-flavour, and then be swallowed and forgotten" (MP, p. 344). Perhaps Henry simply views others' preferences as long-lasting as his own; at one moment Henry "wished he had been a William Price, distinguishing himself and working his way to fortune and consequence with so much self-respect and happy ardour" but at the next "found it was as well to be a man of fortune at once with horses and grooms at his command" (MP, p. 276).

When rationalizations change from one minute to the next, we approach the absurd. When James Morland asks Isabella Thorpe to dance, she says that she cannot while Catherine Morland has no partner: "I would not stand up without your dear sister for all the world; for if I did we should certainly be separated the whole evening." But three minutes later, Isabella tells Catherine, "My dear creature, I am afraid I must leave you, your brother is so amazingly impatient to begin; I know you will not mind my going away" (NA, p. 47). The next day, when the four go on a drive, John Thorpe brags to Catherine about his own carriage, saying that her brother James's carriage is "the most devilish little ricketty business I ever beheld. . . . I would not be bound to go two miles in it for fifty thousand pounds." Taking him seriously, Catherine urges him to warn James of the danger and turn back. Then John Thorpe proclaims that James's "carriage is safe enough, if a man knows how to drive it. . . . I would undertake for five pounds to drive it to York and back again, without losing a nail." Catherine "listened with astonishment...for she had not been brought up...to know to how many idle assertions and impudent falsehoods the excess of vanity will lead. Her own family were plain, matter-of-fact people...not in the habit therefore of telling lies to increase their importance, or of asserting at one moment what they would contradict the next" (NA, pp. 61–62). John Thorpe's boasts are not mere logical contradictions such as "Everything I say is false"; one can make contradictory statements without changing one's preferences. What makes John Thorpe absurd is the flipping of his stated preferences. Edmund Bertram is aware of this. After first refusing to act in *Lovers' Vows*, he decides that he must take part in order to preserve Mary Crawford from acting an intimate part with a complete stranger, and states, "No man can like being driven into the *appearance* of such inconsistency. After being known to oppose the scheme from the beginning, there is absurdity in the face of my joining them *now*" (MP, pp. 180–81).

Austen tracks how preferences change, but surely the deeper question is how preferences are formed in the first place. Most game theorists, however, consider this question outside the scope of their inquiry; "the classic economic/game theory model...views preferences as exogenous; they are taken as given, and the analysis begins at that point" (Legro 1996, p. 119). Austen is no different. Gratitude explains Henry Tilney's love for Catherine, Elizabeth's love for Mr. Darcy, and also perhaps Fanny's love for Edmund Bertram (after Edmund helps ten-year-old Fanny write a letter to her brother William, "her countenance and a few artless words fully conveyed all their gratitude and delight" [MP, p. 18]), but Austen does not feel it necessary to explain the origins of all loves. For example, Catherine's interest in Henry Tilney is taken for granted, and the beginning of Elinor's connection to Edward Ferrars is described cursorily as "a particular circumstance," "a growing attachment" (SS, p. 17). Edmund's attraction to Mary Crawford begins embarrassingly routinely: "A young woman, pretty, lively, with a harp as elegant as herself; and both placed near a window, cut down to the ground, and opening on a little lawn, surrounded by shrubs in the rich foliage of summer, was enough to catch any man's heart" (MP, p. 76). Mary Crawford's response is equally mundane: "he pleased her for the present; she liked to have him near her; it was enough" (MP, p. 77).

Austen does note that people can acquire new preferences; after Catherine exclaims that Eleanor Tilney has taught her how to love a hyacinth, Henry Tilney remarks, "You have gained a new source of enjoyment, and it is well to have as many holds upon happiness as possible....I am pleased that you have learnt to love a hyacinth. The mere habit of learning to love is the thing" (NA, pp. 178–79). One can also say that Emma and Mr. Knightley illustrate how sometimes you don't know your own preferences until you are forced into action, by a rival for example. Still, Austen is more interested in people's actions once they have their preferences.

CONSTANCY

The opposite of preference change is constancy, which Austen considers a virtue in itself. Austen prizes constancy but distinguishes carefully between constancy and stubbornness, and also clarifies the surprisingly subtle distinction between constancy and inconstancy. Austen understands constancy as fundamentally a strategic process.

Austen's exemplars of constancy are Anne Elliot and Captain Wentworth, whose love for each other remains after a separation of eight years. Constancy is essential not only in the fundamentals of their love but also in its tactical achievement: it is Anne's expression of woman's superior constancy, egged on by Captain Harville, which finally makes Captain Wentworth redeclare his feelings.

Their trajectory analyzes, as well as celebrates, constancy. When she was nineteen, Anne allowed Lady Russell to persuade her to refuse Captain Wentworth's proposal, and he returns eight years later obsessed with the virtue of decisiveness: "He had not forgiven Anne Elliot. She had used him ill; deserted and disappointed him; and worse, she had shewn a feebleness of character in doing so.... She had given him up to oblige others. It had been the effect of over-persuasion. It had been weakness and timidity" (P, p. 66). Walking with Anne's potential rival, Louisa Musgrove, Captain Wentworth declares, "It is the worst evil of too yielding and indecisive a character, that no influence over it can be depended on.... Everybody may sway it.... My first wish for all, whom I am interested in, is that they should be firm" (P, p. 94). But when Louisa falls on her head at Lyme, Anne reflects that her fall was caused by her willful insistence to jump down the steps of the Cobb a second time, even after Captain Wentworth tried to persuade her against it. Anne therefore thinks that Captain Wentworth might question "his own previous opinion as to the universal felicity and advantage of firmness of character" (P, p. 126). Comparing Anne and Louisa, Captain Wentworth learns the correct conceptual distinction: "he had learnt to distinguish between the steadiness of principle and the obstinacy of self-will, between the darings of heedlessness and the resolution of a collected mind" (P, p. 263). Constancy is not about heedlessness and obstinacy, decisiveness for its own sake. Constancy is about steadiness and resolution.

For Anne and Captain Wentworth, constancy requires hard work, patience, and faith, as well as strategically figuring out what the other is thinking: in moments of doubt, Anne relies "on this argument of rational dependance—'Surely, if there be constant attachment on each side, our hearts must understand each other ere long. We are not boy and girl, to be captiously irritable, misled by every moment's inadvertence, and wantonly playing with our own happiness' " (P, p. 240). In contrast, as mentioned in chapter 5, when Louisa jumps off the steps of the Cobb, she does so half a second too early, before Captain Wentworth is ready to hold her: "he put out his hands; she was too precipitate by half a second, she fell on the pavement on the Lower Cobb, and was taken up lifeless!" (P, p. 118). Louisa's mistake is not just willfulness, but a lack of strategic thinking; she does not think about Captain Wentworth's mental state before jumping and simply assumes he is ready. Louisa acts before she and Captain Wentworth understand each other. Like Fox, who superficially imitates Rabbit's fish-stealing technique only to be struck on the head by the fisherman, Louisa tries to attract Captain Wentworth with superficial displays of decisiveness (and is also struck on the head), without understanding what constancy truly requires.

True constancy requires strategic thinking. After they finally understand their love for each other, Captain Wentworth explains to Anne how discouraged he was to see her the target of her cousin Mr. Elliot and in the company of Lady Russell: "[T]o see your cousin close by you, conversing and smiling, and feel all the horrible eligibilities and proprieties of the match! To consider it as the certain wish of every being who could hope to influence you! . . . [W]as not the recollection of what had been, the knowledge of her influence, the indelible, immoveable impression of what persuasion had once done—was it not all against me?" Anne gently replies, "If I was wrong in yielding to persuasion once, remember that it was to persuasion exerted on the side of safety, not of risk. When I yielded, I thought it was to duty; but no duty could be called in aid here. In marrying a man indifferent to me, all risk would have been incurred, and all duty violated" (P, p. 266). Anne's point is that Captain Wentworth, instead of seeing only reminders of how she was earlier persuaded, should realize that she makes her own decisions and should consider the factors involved in her preferences.

Later that day, thinking further, Anne continues, "To me, [Lady Russell] was in the place of a parent. . . . I am not saying that she did not err in her advice. . . . But I mean, that I was right in submitting to her, and that if I had done otherwise, I should have suffered more in continuing the engagement than I did even in giving it up, because I should have suffered in my conscience" (P, pp. 267–68). Captain Wentworth replies with a more relevant and painful counterfactual: "Tell me if, when I returned to England in the year eight, with a few thousand pounds, and was posted into the Laconia, if I had then written to you, would you have answered my letter? would you, in short, have renewed the engagement then?" Anne, of course, would have. Captain Wentworth admits, "I was proud, too proud to ask again. I did not understand you. I shut my eyes, and would not understand you, or do you justice.... Six years of separation and suffering might have been spared" (P, p. 268). Obsessed with what he considers Anne's persuadability, purposefully trying not to understand her, not considering whether Anne's desire to marry him might endure while perceived risks fade, Captain Wentworth acts in a manner even more inconstant.

As for inconstancy, Austen playfully considers how it is not necessarily a well-defined concept. Frank Churchill's visits to Highbury are unpredictable because he must be given leave by his adoptive mother, Mrs. Churchill. Emma complains, "[I]t seems to depend upon nothing but the ill-humour of Mrs. Churchill, which I imagine to be the most certain thing in the world." But Mrs. Weston replies, "My Emma!...[W]hat is the certainty of caprice?" (E, p. 131). Is it possible to be constantly inconstant? Does being inconstantly inconstant mean that you are sometimes constant? After Mr. Bingley tells Mrs. Bennet that "if I should resolve to quit Netherfield, I should probably be off in five minutes," Mr. Darcy faults him for seeming to take pride in his own ductility (PP, p. 46). As mentioned earlier in this chapter, Mr. Darcy suggests that even if Mr. Bingley were on his horse ready to leave, if a friend told him that he should stay, he would just as easily decide to stay for another month. Mr. Bingley jokes that Mr. Darcy would rather him "ride off as fast as I could," and Elizabeth cannot help remarking, "Would Mr. Darcy then consider the rashness of your original intention as atoned for by your obstinacy in adhering to it?" (PP, p. 54). Elster (1983, p. 11) similarly notes, "[T]rying to be spontaneous is a self-defeating plan, since the very act of trying will interfere with the goal." Mr. Bingley had been discouraged by Mr. Darcy's opinion that Jane did not return his affection, but is easily reconvinced when Mr. Darcy assures him that she really does love him. "Elizabeth could not help smiling at his easy manner of directing his friend," and indeed in so easily returning to his original position, it is not clear whether Mr. Bingley should be valued for being ductile or steadfast (PP, p. 411).

After recovering from her illness, Marianne's resolution to become more studious is driven by a zeal as unflagging as ever: "Elinor honoured her for a plan which originated so nobly as this; though smiling to see the same eager fancy which had been leading her to the extreme of languid indolence and selfish repining, now at work in introducing excess into a scheme of such rational employment and virtuous self-control"

(SS, p. 389). In trying to change her own character so abruptly, Marianne acts in a way deeply consistent with her character; for her, changing herself slowly and judiciously, becoming self-controlled in a self-controlled manner, would be more truly inconstant. The unrestrained eagerness that had attached her so rapidly and miserably with Willoughby serves in good stead when she marries Colonel Brandon: "Marianne could never love by halves; and her whole heart became, in time, as much devoted to her husband, as it had once been to Willoughby" (SS, p. 430).

Whether Marianne's eagerness makes her inconstant or constant depends on when the question is asked: when she switches her affections from Willoughby to Colonel Brandon, it makes her inconstant, but once she is attached to Colonel Brandon, it makes her constant. When Mr. Bingley is persuaded to drop Jane, his guidability makes him inconstant, but when he is reconvinced, it makes him constant. When should a person's constancy be evaluated? Only at the end, according to Elizabeth Bennet. After Elizabeth and Mr. Darcy understand their affection for each other, she tells him that their previous feelings should be entirely discounted; their feelings "are now so widely different from what they were then, that every unpleasant circumstance attending it, ought to be forgotten. You must learn some of my philosophy. Think only of the past as its remembrance gives you pleasure" (PP, p. 409).

CHAPTER TEN

Austen on Strategic Thinking's Disadvantages

NO ANALYSIS OF strategic thinking would be complete without recognizing its costs and disadvantages. Game theorists rarely go this far, but Austen's ambition is comprehensive.

First and most obviously, Austen notes that strategic thinking takes mental effort: one's strategic thinking capacity is not infinite and strategic thinking competes with other cognitive demands. Elizabeth early on misperceives Mr. Darcy partly because she devotes all of her detection efforts toward Mr. Bingley: "Occupied in observing Mr. Bingley's attentions to her sister, Elizabeth was far from suspecting that she was herself becoming an object of some interest in the eyes of his friend" (PP, p. 25). After Mr. Darcy proposes, it takes effort for Elizabeth to hold the news inside and carefully strategize about how to tell her sister Jane without disappointing her: "It was not without an effort meanwhile that she could wait even for Longbourn, before she told her sister of Mr. Darcy's proposals.... [She felt] such a temptation to openness as nothing could have conquered, but the state of indecision in which she remained, as to the extent of what she should communicate; and her fear ... of being hurried into repeating something of Bingley, which might only grieve her sister farther" (PP, p. 241). When Mr. Bingley visits the Bennets with Mr. Darcy after a long absence, "Jane was anxious that no difference should be perceived in her at all, and was really persuaded that she talked as much as ever. But her mind was so busily engaged, that she did not always know when she was silent"; the task of playing it cool is so demanding that Jane cannot monitor her own conversation (PP, p. 373).

After Anne talks to Captain Wentworth at a concert in Bath, his "sentences begun which he could not finish—his half averted eyes, and more than half expressive glance,—all, all declared that he had a heart returning to her at least.... He must love her. These were thoughts, with their attendant visions, which occupied and flurried her too much to leave her any power of observation; and she passed along the room without having a glimpse of him, without even trying to discern him. When their places were determined on, and they were all properly arranged, she looked round to see if he should happen to be in the same part of the room, but he was not; her eye could not reach him." Anne's occupation in interpreting Captain Wentworth's feelings and envisioning their implications has real costs in that she loses the chance to sit near him. But Anne accepts that her strategic thinking capacity is not unbounded, and that "she must consent for a time to be happy in a humbler way" (P, p. 202). Note that Captain Wentworth's feelings are themselves indicated by his own cognitive difficulty in forming complete sentences. Mr. Woodhouse's statement that "Emma never thinks of herself, if she can do good to others" is intended to be a testament to Emma's kindness, but is more accurately a statement that Emma spends so much time manipulating others that she does not think about her own wants (E, p. 12).

If you are known to have strategic skills, you can be overburdened by others' requests. When Anne stays with her sister Mary, she is burdened with "too much confidence by all parties.... Known to have some influence with her sister, she was continually requested, or at least receiving hints to exert it, beyond what was practicable. 'I wish you could persuade Mary not to be always fancying herself ill,' was Charles's language; and, in an unhappy mood, thus spoke Mary;—'I do believe if Charles were to see me dying, he would not think there was anything the matter with me. I am sure, Anne, if you would, you might persuade him that I really am very ill' " (P, p. 48).

Another cost of strategic thinking is a more complicated moral life. Elinor has to do all the lying because Marianne does not want to: "it was impossible for [Marianne] to say what she did not feel, however trivial the occasion; and upon Elinor therefore the whole task of telling lies when politeness required it, always fell" (SS, p. 141). Elinor must pay the cost, in both effort and honesty, that Marianne cannot bear. Less trivially, Jane Fairfax admits that the "consciousness of having done amiss, had exposed her to a thousand inquietudes, and made her captious and irritable" during her secret engagement (E, p. 457). If Jane Fairfax were more like Marianne and could not help bursting out, her hopes of marrying Frank Churchill might have been endangered, but at least she would not have had to endure a "life of deceit" (E, p. 501).

Being good at strategic thinking enlarges the scope of regret, as one considers oneself responsible for a wider range of outcomes. When Emma is shocked to hear that Mr. Knightley might be interested in Harriet Smith, "if Harriet were to be the chosen... what could be increasing Emma's wretchedness but the reflection never far distant from her mind, that it had been all her own work?" (E, p. 460).

Strategic thinking can be morally compromising, as it makes one more adept at creating excuses for others' behavior. Emma tells Mr. Knightley that they should excuse Frank Churchill's long delay in paying his respects to his new mother-in-law, Mrs. Weston, because of the difficulty of releasing himself from his adopted family: "I wish you would try to understand what an amiable young man may be likely to feel in directly opposing those, whom as child and boy he has been looking up to all his life" (E, p. 159). When Fanny warns Edmund that Henry Crawford seems to be interested more in the engaged Maria than in the unattached Julia, Edmund explains it away by saying, "I believe it often happens, that a man, before he has quite made up his own mind, will distinguish the sister or intimate friend of the woman he is really thinking of, more than the woman herself" (MP, pp. 136–37).

Simply having strategic insights present in your mind can be painful, regardless of how useful they are. Elizabeth tries to convince Jane that Caroline Bingley and Mr. Darcy are trying to prevent Mr. Bingley from marrying her, and Jane replies, "I have no idea of there being so much design in the world as some persons imagine.... Do not distress me by the idea" (PP, pp. 154–55). Jane might be in denial, but even someone as hard-headed as Elizabeth would not wish upon anyone what Fanny has to go through when she must help Edmund and Mary Crawford rehearse *Lovers' Vows* together, and thereby contemplate what their rehearsal might lead to. Fanny, prompting Edmund and Mary from the script, is forced to "feel too much.... [A]gitated by the increasing spirit of Edmund's manner, [Fanny] had once closed the page and turned away exactly as he wanted help. It was imputed to very reasonable weariness, and she was thanked and pitied; but she deserved their pity more than she hoped they would ever surmise" (MP, pp. 199–200).

In some situations, it is better to be unstrategic and not think too much about what others will do. Mr. Collins looks for a bride in sequential order: he plans to propose to Jane first, he proposes to Elizabeth next, and finally succeeds with Charlotte Lucas. His algorithm is quite unstrategic in that he does not think much about how each target will react, for example, how she might feel knowing that she is a second or third choice. But Mr. Collins's method is admirably direct, and compares favorably to the haphazardness of Elizabeth and Mr. Darcy and the hesitancy of Anne and Captain Wentworth.

Being good at strategic thinking can keep people from helping you. For her sister's and mother's sake, Elinor holds inside the secret of Lucy's engagement and pretends as if she doesn't care about Edward. When Elinor, Marianne, and Mrs. Dashwood learn that Lucy has become Mrs. Ferrars, and therefore conclude (as it turns out, erroneously) that she married Edward, Mrs. Dashwood "was shocked to perceive by Elinor's countenance how much she really suffered" and "found that she had been misled by the careful, the considerate attention of her daughter. ... She feared that under this persuasion she had been unjust, inattentive, nay, almost unkind, to her Elinor" (SS, pp. 400, 403). Had Elinor been less skillful, she could have received her mother's comfort earlier.

Strategic thinking is not the same as craftiness or cunning, but it is not charming. When Catherine breaks her date with the Tilneys to go on a walk, she is not entirely at fault, since the rain made the original plan uncertain and the Tilneys did not show up precisely on time. But she speeds toward Henry Tilney to apologize: "Feelings rather natural than heroic possessed her ... [I]nstead of proudly resolving ... to leave to him all the trouble of seeking an explanation ... she took to herself all the shame of misconduct, or at least of its appearance, and was only eager for an opportunity of explaining its cause" (NA, p. 92). Catherine could have tried to get Henry to make the first move, but instead she goes for what is natural, and this lack of strategicness is winning. Elizabeth teases her sister Jane: "With *your* good sense, to be so honestly blind to the follies and nonsense of others! Affectation of candour is common enough.... But to be candid without ostentation or design ... belongs to you alone" (PP, p. 16). Elizabeth, like Flossie Finley, knows that what seems like naivety can itself be a strategy, and prizes Jane for being genuinely nonstrategic. When Mr. Bingley jokes, "[T]o be so easily seen through I am afraid is pitiful," Elizabeth assures him that "[i]t does not follow that a deep, intricate character is more or less estimable than such a one as yours" (PP, p. 46).

Why is artlessness charming? Austen observes that for men, "imbecility in females is a great enhancement of their personal charms" (NA, p. 112) and a woman's lack of agency can be attractive in itself, for example when Henry Crawford falls in love with Fanny. Emma, reflecting to herself, has a similar attitude toward Harriet: "Warmth and tenderness of heart, with an affectionate, open manner, will beat all the clearness of head in the world, for attraction.... Dear Harriet!—I would not change you for the clearest-headed, longest-sighted, best-judging female breathing" (E, pp. 289–90). Emma regards tenderness of heart as the opposite of clear thinking, farsightedness, and good judgment, and prizes Harriet for being easily knowable and guidable. For people like Henry Crawford and Emma, artless people are convenient because they can be put in any social role one wants, such as wife, without having to think about what they want or what they will do.

Artlessness can also be appreciated as sincerity. Anne "could so much more depend upon the sincerity of those who sometimes looked or said a careless or a hasty thing, than of those whose presence of mind never varied, whose tongue never slipped" (P, p. 175). A strategic action is not necessarily insincere. For example, when Elinor holds in Lucy's secret, she might not be open with Marianne but she is not insincere; if anything, she is sincerely withholding the truth. But a clearly unstrategic action is sincere in the sense that there are obviously no hidden motivations.

When Catherine meets Eleanor Tilney for the second time, " 'How well your brother dances!' was an artless exclamation of Catherine's towards the close of their conversation, which at once surprised and amused her companion" (NA, p. 70). After Sir Thomas comes home and discovers *Lovers' Vows* in production, Tom Bertram and Mr. Yates try to spin their way out, saying that Sir Thomas himself had "so often encouraged the sort of thing in us formerly." But the dullard Mr. Rushworth pronounces, "I am not so fond of acting as I was at first. I think we are a great deal better employed, sitting comfortably here among ourselves, and doing nothing," and Sir Thomas heartily approves (MP, pp. 216–18). After Tom's and Mr. Yates's slickness, Mr. Rushworth's painful sincerity is, for once, well timed, and "[i]t was impossible for many of the others not to smile" (MP, p. 218). When Emma meets Frank Churchill for the first time, she knows that Frank's father, Mr. Weston, has hopes that the two might marry: "She had no doubt of what Mr. Weston was often thinking about. His quick eye she detected again and again glancing towards them with a happy expression." In comparison, Mr. Woodhouse is comfortably clueless: "Her own father's perfect exemption from any thought of the kind, the entire deficiency in him of all such sort of penetration or suspicion, was a most comfortable circumstance. . . . She blessed the favouring blindness" (E, pp. 207–208).

If others do not think you are strategic, then they confide in you, thinking that you cannot possibly be leading them on. Before Mary Crawford leaves for London, she checks in with Fanny about Edmund's interest: "she had hoped to hear some pleasant assurance of her power, from one who she thought must know" (MP, p. 335). Edmund is visiting Mr. Owen, who has marriageable sisters, and when Mary asks, "[P]erhaps you do not think him likely to marry at all—or not at present," she is relieved to hear Fanny reply, "No, I do not" (MP, p. 336). Mary does not realize that Fanny has her own interest in Edmund's singleness, and that she is revealing to Fanny her desperation and giving Fanny the opportunity to manipulate her expectations. In answering that Edmund will not soon marry, Fanny "hop[es] she did not err either in the belief or the acknowledgment of it"; Fanny is aware that her answer is not just a matter of factual correctness but has strategic implications. If Fanny says that Mary need not worry about the Owen sisters, Mary will think Edmund more truly attached to her, and any encouragement might make Edmund and Mary more likely to marry; on the other hand, if Fanny says that Edmund is at risk from the Owen sisters, then Mary might make a strong preemptive move toward Edmund, which for Fanny might be even worse. Fanny makes her decision and hopes for the best. Mary, perhaps suspecting Fanny's strategicness, "looked at her keenly;

and gather[ed] greater spirit from the blush soon produced from such a look" (MP, p. 336). Mary is assured by Fanny's blush, which seems to guarantee sincerity.

Similarly, if others think you are strategic, they do not confide in you because they think you already know everything. Mrs. Weston worries that Emma must be severely disappointed about the engagement of Frank Churchill and Jane Fairfax, and thus hesitates telling her directly: "Have you indeed no idea? . . . Cannot you, my dear Emma—cannot you form a guess as to what you are to hear?" (E, p. 430). Harriet also asks Emma, "Had you any idea . . . of his being in love with her?—You, perhaps, might.—You (blushing as she spoke) who can see into everybody's heart" (E, p. 441). Frank Churchill writes to Mrs. Weston that he should not be blamed for his casual familiarity with Emma because Emma must have known that he was already engaged: "I remember that I was within a moment of confessing the truth, and I then fancied she was not without suspicion; but I have no doubt of her having since detected me, at least in some degree.—She may not have surmised the whole, but her quickness must have penetrated a part" (E, p. 478). These examples are trivial, however, compared to the misunderstanding that follows Mr. Knightley's admission that he envies Frank Churchill "in one respect." Emma thinks that "[t]hey seemed to be within half a sentence of Harriet" and pauses for a moment, trying to figure out how to change the subject, but Mr. Knightley interrupts, "You will not ask me what is the point of envy.—You are determined, I see, to have no curiosity.—You are wise— but I cannot be wise. Emma, I must tell you what you will not ask, though I may wish it unsaid the next moment." Mr. Knightley knows that Emma is strategically skilled, and thus when Emma does not ask what exactly he envies, he concludes that she understands the possibility of his affection but does not want to receive it. Since he knows that Emma is a quick thinker, he takes her pause as an expression of indifference; such a pause from a slower-thinking person would mean nothing. Emma compounds the error by replying, "Oh! then, don't speak it, don't speak it. . . . Take a little time, consider, do not commit yourself" (E, p. 467). Luckily, Emma reconsiders and continues the conversation, averting disaster. For Mr. Knightley, to be sincere one cannot be wise.

Being strategically skilled can you make you see strategicness where none exists. As mentioned in chapter 5, when Mrs. Jennings hands Marianne a letter from her mother, Marianne is severely disappointed that it is not from Willoughby, and angrily blames Mrs. Jennings for intentionally harming her, even though "Mrs. Jennings was governed in it by an impulse of the utmost good-will" (SS, p. 229). When Elizabeth visits the Collins residence after Charlotte's marriage, she thinks that Mr. Collins is trying to impress her with the lifestyle she could have had if she had accepted his proposal: "she could not help fancying that in displaying the good proportion of the room, its aspect and its furniture, he addressed himself particularly to her, as if wishing to make her feel what she had lost in refusing him" (PP, p. 177). This thought is not unreasonable, but later when Elizabeth and Charlotte sit together in the house, "Elizabeth at first had rather wondered that Charlotte should not prefer the dining parlour for common use; it was a better sized room, and had a pleasanter aspect; but she soon saw that her friend had an excellent reason for what she did, for Mr. Collins would undoubtedly have been much less in his own apartment, had they sat in one equally lively; and she gave Charlotte credit for the arrangement" (PP, p. 189). This conclusion, that Charlotte purposefully sits in an unattractive room to avoid Mr. Collins, goes too far. Elizabeth has no reason to believe, for example, that any difference in pleasantness even registers with Mr. Collins, who is hardly an aesthete. Elizabeth, who wants to believe that she and Charlotte are still close, is trying too hard to appreciate Charlotte's strategicness.

Being good at strategic thinking can make you proud. Emma is, of course, the best example: after Mrs. Weston tells her that she thinks that Mr. Knightley likes Jane Fairfax, Emma maneuvers Mr. Knightley into saying that he has no interest in Jane, and afterward Emma "triumphantly" gloats: "Well, Mrs. Weston...what do you say now to Mr. Knightley's marrying Jane Fairfax?" Mrs. Weston replies, "Do not beat me" (E, p. 312). Another example is Fanny Dashwood, who, "rejoicing in her escape, and proud of the ready wit that had procured it," disastrously invites Lucy and Anne Steele to her home to prevent her husband from inviting Elinor and Marianne (SS, p. 287). Not quite as blatantly, Edmund Bertram tells Fanny Price that Henry Crawford should have consulted him before proposing to her: "I wish he had known you as well as I do, Fanny. Between us, I think we should have won you. My theoretical and his practical knowledge together, could not have failed. He should have worked upon my plans" (MP, p. 402). Edmund, however wrong, thinks that Henry should have acknowledged his superior theoretical (strategic) knowledge and personal knowledge of Fanny. Similarly, Henry Tilney "believes that he knows women's minds better than they do" (Johnson 1988, p. 37). After Catherine tells Henry and Eleanor Tilney that Isabella broke her engagement with Catherine's brother James, Henry sounds like he is trying to win a mind-reading contest: "You feel, I suppose, that in losing Isabella, you lose half yourself: you feel a void in your heart which nothing else can occupy. Society is becoming irksome; and as for the amusements in which you were wont to share at Bath, the very idea of them without her is abhorrent. You would not, for instance, now go to a ball for the world.

You feel that you have no longer any friend to whom you can speak with unreserve; on whose regard you can place dependence; or whose counsel, in any difficulty, you could rely on. You feel all this?" Catherine responds, "No...I do not—ought I?" (NA, pp. 212–13).

The first step in strategic thinking is to realize that other people can think differently from you. But when you are overconfident in your own strategic ability, you think that other people are so transparent that you again confuse your idea of what they are thinking with what they are in fact thinking. According to Schulz (2010, p. 331), "*Pride and Prejudice* is a book about people who, believing themselves to be astute scholars of human nature, persistently and dramatically misunderstand each other." Elizabeth wants to think that Charlotte is still close to her and obviously shares her preference to avoid Mr. Collins, Edmund wants to think that he knows Fanny best, and Henry Tilney, who when first meeting Catherine is willing to predict exactly what she will write in her journal about him, wants to enhance his status as telepathic tutor. Of course, Elizabeth soon learns the extent to which she can misunderstand others (notably Mr. Darcy), Edmund learns from Mary Crawford how wrong his estimation of others can be, and Henry Tilney learns that taking Catherine's concerns (which she says out loud, telepathy being unnecessary) for granted leaves her at the mercy of his father. "[B]lindness is the reward of assuming a godlike control" (Knox-Shaw 2004, pp. 199–200). True strategic wisdom is not proud.

CHAPTER ELEVEN

Austen's Intentions

ONE MIGHT SAY that strategic thinking is so integral to human interaction that it cannot be avoided; indeed, any narrative in which a character anticipates another's actions illustrates strategic thinking to some degree. But illustrating strategic thinking is one thing; making it a central theoretical concern is altogether more ambitious.

Is this Austen's intention? If not, one would have to explain the inclusion of many particular and otherwise unnecessary details, such as Elizabeth's argument to Jane that the pain of upsetting Mr. Bingley's sisters relative to the joy of marrying him is best measured by whether Jane chooses to refuse him, or Mrs. Weston's argument that Jane Fairfax's decision to spend time with Mrs. Elton is explained by the tiresomeness of the alternative, staying at home with her aunt. One would have to explain Austen's superabundance of "schemes" and prizing of "penetration" (according to Vermeule [2010, p. 187], "a word Austen plays on obsessively"). One would have to explain Austen's fairly direct theoretical statements, such as Elinor's doctrine that others should influence only your behavior, not your understanding, or Henry Tilney's telling Catherine that she thinks of others' actions in terms of her own motivations.

The question of intention is not essential for Austen in any case. General Tilney first invites and later expels Catherine Morland from Northanger Abbey; these two actions are taken with completely opposite intentions, but both create opportunities for Catherine to better secure Henry Tilney's heart. Lady Catherine orders Elizabeth Bennet to promise not to marry Mr. Darcy, but thereby unintentionally creates an opportunity for Elizabeth to send a message to him. Does Elizabeth anticipate that Lady Catherine will tell Mr. Darcy that she refused to promise not to marry him? Does Elizabeth rebuff Lady Catherine with the explicit intention of giving hope to Mr. Darcy? We do not know, but Elizabeth loses little by trying. We know that Sir Thomas sends Fanny Price to Portsmouth with the intention of getting her to marry Henry Crawford and Mrs. Allen takes Catherine Morland to Bath with no intentions whatsoever. Does Mrs. Gardiner bring Elizabeth to Pemberley with the intention of possibly running into Mr. Darcy? Does Mrs. Croft persuade her husband, Admiral Croft, to rent Kellynch Hall with the intention of bringing her brother Captain Wentworth into Anne Elliot's vicinity?

We have no direct evidence on these questions, but it does not matter much; what is important is what our heroines do once given these opportunities. Regardless of whether Austen intends to impart game theory in her novels, it is up to the reader to receive it.

The most specific "smoking gun" evidence that Austen is centrally concerned with strategic thinking is how she employs children: when a child appears, it is almost always in a strategic context. Children are often pawns or bit players in an adult's strategic actions. For example, when the Dashwoods meet Lady Middleton for the first time, "Lady Middleton had taken the wise precaution of bringing with her their eldest child, a fine little boy about six years old, by which means there was one subject always to be recurred to by the ladies in case of extremity.... On every formal visit a child ought to be of the party, by way of provision for discourse" (SS, pp. 36–37). When the Steele sisters want to suck up to Lady Middleton, they know that their best move is to praise her children: "a fond mother... in pursuit of praise for her children... will swallow any thing; and the excessive affection and endurance of the Miss Steeles towards her offspring, were viewed therefore by Lady Middleton without the smallest surprise or distrust" (SS, p. 139). Anne's nephew Charles conveniently falls and dislocates his collarbone, allowing her to stay home to take care of him and thereby delay the anxiety of seeing Captain Wentworth after eight years apart. Later her two-year-old nephew Walter attaches himself to her back, enabling Captain Wentworth to wordlessly show affection to her by lifting him off. After their disagreement over whether Robert Martin is of sufficient stature to wed Harriet Smith, Emma uses her eight-month-old niece to reconcile with Mr. Knightley: "she hoped it might rather assist the restoration of friendship, that when he came into the room she had one of the children with her" (E, p. 105).

Children are also brought in as students of strategic thinking themselves. As discussed in chapter 5, the younger sisters Kitty Bennet, Margaret Dashwood, and Sarah Morland start laughably unsophisticated but by the end learn. For example, after Kitty Bennet tells her father that she will not run away like Lydia, Mr. Bennet, who cannot stop joking even in a desperate situation, exclaims, "No, Kitty, I have at last learnt to be cautious, and you will feel the effects of it. No officer is ever to enter into my house again, nor even to pass through the village" and Kitty, "who took all these threats in a serious light, began to cry" (PP, pp. 330, 331). But later when Mr. Bingley shows up unexpectedly early for a dinner appointment at the Bennet residence, making Mrs. Bennet hurry her daughters' hair and dress preparations, Jane notes that "Kitty is forwarder than either of us, for she went up stairs half an hour ago" (PP, p. 381). Kitty is not just a "surplus" character, "a representation of how we perceive or constitute people at the periphery" (Woloch 2003, pp. 118–19); Kitty, like Margaret Dashwood, Sarah Morland, and Susan Price, is a Younger Sister Who Learns Strategic Thinking. When James Morland is engaged to Isabella Thorpe, Isabella's younger sisters Anne and Maria are expected to figure it out for themselves. When Mr. Knightley is engaged to Emma, Mr. Knightley wonders why his brother does not seem entirely surprised, and guesses, "I dare say there was a difference when I was staying with them the other day. I believe I did not play with the children quite so much as usual. I remember one evening the poor boys saying, 'Uncle seems always tired now' " (E, pp. 507–8). Thus the task of sensing engagements is imputed to children.

A child's strategizing begins the Dashwoods' plight: Elinor and Marianne's father, Mr. Henry Dashwood, had been the heir of his uncle, and had depended on this inheritance for the maintenance of his wife and daughters; however, this uncle decided to leave Norland wholly to the four-year-old son of Henry Dashwood's son John. This revision was accomplished "by such attractions as are by no means unusual in children of two or three years old; an imperfect articulation, an earnest desire of having his own way, many cunning tricks, and a great deal of noise" (SS, p. 5). But the birth of a child also initiates a chain of decisions that in the end brings Elinor victory: to take care of her new grandchild, Mrs. Jennings leaves Marianne and Elinor alone in London; John Dashwood thus suggests to his wife that his sisters should stay with them; his wife, Fanny Dashwood, thus invites instead Lucy and Anne Steele in order to keep Elinor away from her brother Edward Ferrars; Anne observes the Ferrars family's fondness for Lucy and reveals Lucy and Edward's secret engagement; Edward is thrown off by his family; thus Lucy marries Robert Ferrars and releases Edward, leaving him free to propose to Elinor.

Children learn how to strategize at an early age. Lady Middleton's son John learns how to provoke by "taking Miss Steele's pocket handkerchief, and throwing it out of window," and her three-year-old daughter Annamaria, after she is accidentally scratched by a pin, is showered with attention and sugar plums: "With such a reward for her tears, the child was too wise to cease crying" (SS, pp. 139, 140).

Since learning strategic thinking starts early, lessons missed or mistaught in your childhood can trip you up later. As children, Maria and Julia Bertram ask their aunt Mrs. Norris why Fanny Price is "so odd and so stupid. Do you know, she says she does not want to learn either music or drawing." Mrs. Norris replies that the Bertram sisters should "pity her deficiency.... [I]t is much more desirable that there should be a difference" between them and Fanny (MP, p. 21). Instead of asking the Bertram girls to put themselves in Fanny's place ("Doubtless she was terrified of the prospect of mockery from Julia and Maria," according to Sutherland and Le Faye [2005, p. 167]), Mrs. Norris reinforces their status-oriented thinking. Mr. Darcy tells Elizabeth, "As a child I was taught what was *right*, but I was not taught to correct my temper.... I was spoilt by my parents, who... allowed, encouraged, almost taught me to be selfish and overbearing.... You taught me a lesson, hard indeed at first, but most advantageous" (PP, pp. 409–10). Mr. Darcy's lesson is that a proposal is a strategic situation and a man cannot take a woman's answer for granted: "I came to you without a doubt of my reception. You shewed me how insufficient were all my pretensions" (PP, p. 410).

One can always learn more about strategic thinking, throughout one's life. Emma learns that overstrategizing has its pitfalls, and hopes that "the lessons of her past folly might teach her humility and circumspection in future" (E, p. 519). When Edmund says that he will write Fanny when he has important news, Fanny realizes that he must be talking about his planned proposal to Mary Crawford, and notes how strange it is "[t]hat a letter from Edmund should be a subject of terror!... The vicissitudes of the human mind had not yet been exhausted by her" (MP, p. 431). Even someone as old as Sir Thomas is instructed by time that one's preferences can change as circumstances change: "the high sense of having realised a great acquisition in the promise of Fanny for a daughter, formed just such a contrast with his early opinion on the subject when the poor little girl's coming had been first agitated, as time is for ever producing between the plans and decisions of mortals, for their own instruction, and their neighbour's entertainment" (MP, p. 546). Mr. Darcy says, "I certainly have not the talent which some people possess... of conversing easily with those I have never seen before. I cannot catch their tone of conversation, or appear interested in their concerns." But Elizabeth, seated at the pianoforte, replies that it's about practice, not talent: "My fingers... do not move over this instrument in the masterly manner which I see so many women's do.... But then I have always supposed it to be my own fault—because I will not take the trouble of practising. It is not that I do not believe *my* fingers as capable as any other woman's of superior execution" (PP, p. 197). It is always possible to learn.

Strategic thinking is not part of a standard scholarly education. Anne's friend Mrs. Smith praises her Nurse Rooke: "She is a shrewd, intelligent, sensible woman. Hers is a line for seeing human nature; and she has a fund of good sense and observation which, as a companion, make her infinitely superior to thousands of those who having only received 'the best education in the world,' know nothing worth attending to" (P, pp. 168–69). The traditional scholarly disciplines are no help either.

When Emma hears that Frank Churchill appeared out of nowhere to rescue Harriet Smith from begging gypsies, "a fine young man and a lovely young woman thrown together in such a way, could hardly fail of suggesting certain ideas to the coldest heart and the steadiest brain.... Could a linguist, could a grammarian, could even a mathematician have seen what she did, have witnessed their appearance together, and heard their history of it, without feeling that circumstances had been at work to make them peculiarly interesting to each other?—How much more must an imaginist, like herself, be on fire with speculation and foresight!" (E, p. 362). Strategic thinking is the field of study of the "imaginist," whose academic specialty is no less important than the traditional fields of linguistics, grammar, and mathematics. "Imaginist," coined by Austen, is possibly the first specialized term for game theorist.

The instructional materials for the imaginist are novels, "in which the most thorough knowledge of human nature [and] the happiest delineation of its varieties . . . are conveyed to the world in the best chosen language" (NA, p. 31). Nonfiction is not helpful: after Catherine returns home and mopes over Henry Tilney, her mother uselessly asks her to read an essay from a book "about young girls that have been spoilt for home by great acquaintance" (NA, p. 250). Mary Bennet learns her useless maxims from "great books" on proper conduct (PP, p. 7).

To see that Austen's novels are game theory textbooks, consider how they begin and end. All six novels are set into motion with some sort of strategic manipulation. The cunning of John Dashwood's four-year-old son removes the Dashwood sisters and their mother from Norland, and John Dashwood's commitment to his father's last request that he support them is whittled away by his wife down to "presents of fish and game, and so forth, whenever they are in season" (SS, p. 13). Jane Bennet falls ill and stays at Netherfield because her mother sent her on horseback in the rain. Emma sets up her governess Miss Taylor with Mr. Weston, and decides upon Harriet Smith and Mr. Elton as her next demonstration. Sir Walter Elliot lets out Kellynch Hall because his lawyer Mr. Shepherd finds Admiral Croft. Sir Thomas is convinced by Mrs. Norris to take in Fanny only to find out "with some surprise, that it would be totally out of Mrs. Norris's power to take any share in the personal charge of her" (MP, p. 9); thus Mrs. Norris gets an unpaid "personal assistant" (Sutherland and Le Faye 2005, p. 162). Catherine's lessons start at Bath, but some unnamed entity must put her there: "when a young lady is to be a heroine, the perverseness of forty surrounding families cannot prevent her. Something must and will happen to throw a hero in her way" (NA, p. 9).

The novels periodically include puzzles in strategic thinking. For example, in reviewing how Robert Ferrars and Lucy Steele ended up married, to Elinor's "own heart it was a delightful affair, to her imagination it was even a ridiculous one, but to her reason, her judgment, it was completely a puzzle" (SS, p. 412). When Henry Tilney explains to Catherine why his father kicked her out, Austen leaves the details to the reader: "I leave it to my reader's sagacity to determine how much of all this it was possible for Henry to communicate at this time to Catherine, how much of it he could have learnt from his father, in what points his own conjectures might assist him, and what portion must yet remain to be told in a letter from James" (NA, p. 256). Todd (2006, p. 106) argues that in Austen's novels, "[t]he habit of guessing and puzzling . . . may influence the reader's response and lead to detection of hidden strategies. . . . [and] secret manoeuvrings."

Do the Knightley brothers conspire to bring Harriet Smith and Robert Martin together in London, as Todd suggests? Sutherland and Le Faye (2005, p. 180) write an entire quizbook along these lines, asking, for example, whether Mary Crawford encourages her brother Henry to pursue Fanny in order to get Fanny off the market for Edmund (see also Sutherland 1999 and Mullan 2012). Does Henry Tilney encourage his brother Captain Tilney to break up James Morland and Isabella Thorpe to ensure that "the odious Isabella will never be his sister-in-law" (Sutherland and Le Faye 2005, p. 153)? Does Mr. Shepherd encourage Sir Walter to reduce his expenses (by moving to Bath) so that there will be more left of his estate when Mrs. Clay, Mr. Shepherd's daughter, possibly marries him? Does Mary Musgrove help her sister Anne Elliot marry Captain Wentworth in order to keep her own husband Charles, who had proposed to Anne earlier, from straying (Sutherland and Le Faye 2005, p. 221)? These exercises are left for the reader.

All six novels conclude with puzzles that require just the right manipulation to secure the desired ending. After Catherine and Henry Tilney are engaged, and wait only upon the consent of General Tilney, the question is posed: "what probable circumstance could work upon a temper like the General's?" The solution is for Henry's sister Eleanor to marry a viscount and thus Eleanor "obtained his forgiveness of Henry, and his permission for him 'to be a fool if he liked it!' " (NA, pp. 259–60). Lady Catherine "was extremely indignant on the marriage of her nephew . . . [and] sent him language so very abusive, especially of Elizabeth, that for some time all intercourse was at an end." Thus Elizabeth persuades Mr. Darcy "to overlook the offence, and seek a reconciliation; and, after a little further resistance on the part of his aunt, her resentment gave way" (PP, p. 430). Similarly, even though Edward Ferrars justly feels "neither humble nor penitent" toward his mother, Elinor persuades him to go to his sister Fanny Dashwood and "personally intreat her good offices in his favour," which results in

Edward being readmitted into his own family (SS, pp. 421–22); once this is accomplished, the very last manipulation is to attach Marianne to Colonel Brandon. The more mature Anne and Captain Wentworth do not have to placate older relatives; the interesting question upon their engagement is what happens to Mr. William Elliot, who had been trying to marry Anne to keep Mrs. Clay away from Sir Walter. Mr. Elliot and Mrs. Clay end up together, "and it is now a doubtful point whether his cunning, or hers, may finally carry the day; whether, after preventing her from being the wife of Sir Walter, he may not be wheedled and caressed at last into making her the wife of Sir William" (P, p. 273). There are no objections to Fanny Price's marriage either, but Lady Bertram cannot part with her: "No happiness of son or niece could make her wish the marriage" (MP, p. 546). The answer to this puzzle is Susan, who "remained to supply her place.—Susan became the stationary niece— delighted to be so!...With quickness in understanding the tempers of those she had to deal with...she was soon welcome, and useful to all" (MP, p. 547).

The most challenging puzzle is how Mr. Woodhouse's reliance on Emma can be reconciled with her marrying. This has been a long-standing problem for Emma, who had vowed earlier never to marry because of her father. With patient "continual repetition," Mr. Wood-house "began to think that some time or other—in another year or two, perhaps—it might not be so very bad if the marriage did take place" (E, pp. 509–10). But the real solution is the arrival of poultry thieves in the neighborhood, which rattles Mr. Woodhouse excessively and makes him eager for the protection of Mr. Knightley as resident son-in-law. Thus Mr. Woodhouse's fretfulness finally pays off. This "accidental" manipulation comes out of nowhere, like the gypsies who allow Frank Churchill to rescue Harriet, but it is not implausible. Austen shows us that seemingly impossible situations can be overcome with just the right change of circumstances, and what seems like a disadvantage, even the entire reason for the problem in the first place, can be used to one's advantage. A successful manipulation is always possible if you are creative enough. Maybe the thievery was just a rumor created by Emma or her confederates, as one of the poultry houses reportedly robbed belonged to Mrs. Weston, Emma's closest friend. Maybe Austen is showing off her own strategic thinking skills: the problem is posed at the very beginning of the novel, remains unchanged throughout, and the reader is given plenty of time to think of a solution. For Austen it is easy.

Anyone can be good (or bad) at strategic thinking, and anyone can learn. But according to Austen, at least some aspects of strategic thinking are the particular domain of women. For example, Catherine's trying to avoid John but attract Henry at a ball is a situation that "every young lady has at some time or other known.... All have been, or at least all have believed themselves to be, in danger from the pursuit of some one whom they wished to avoid; and all have been anxious for the attentions of some one whom they wished to please" (NA, p. 72). When Mr. Knightley proposes, Emma effortlessly draws upon ladies' shared strategic culture: "Her way was clear, though not quite smooth.—She spoke then, on being so entreated.—What did she say?—Just what she ought, of course. A lady always does.—She said enough to show there need not be despair—and to invite him to say more himself" (E, p. 470).

There also seems to be an unspoken pact among women to respect each other's strategic abilities and not use them against each other. When Edmund tells Fanny that Mary Crawford does not understand how Fanny could refuse her brother Henry's proposal, and that "[s]he desires the connection as warmly as your uncle or myself," Fanny replies, "I *should* have thought... that every woman must have felt the possibility of a man's not being approved, not being loved by some one of her sex at least, let him be ever so generally agreeable" (MP, pp. 406, 408). A woman's right to refuse a proposal is something all women should uphold, and Fanny is particularly irked that fellow woman Mary Crawford is willing to compromise on that right just to curry favor with a man. When Emma tells Frank Churchill her suspicions that the pianoforte was sent to Jane Fairfax as a gesture of affection by the married Mr. Dixon, "[s]he doubted whether she had not transgressed the duty of woman by woman, in betraying her suspicions of Jane Fairfax's feelings" (E, p. 249). Caroline Bingley tells Mr. Darcy that Elizabeth "is one of those young ladies who seek to recommend themselves to the other sex by undervaluing their own.... [I]n my opinion, it is a paltry device, a very mean art" (PP, p. 43). But of course in demeaning Elizabeth to Mr. Darcy, Caroline's charge that Elizabeth lacks solidarity applies better to herself. Mr. Darcy understands what she is doing and replies, "Undoubtedly... there is a meanness in *all* the arts which ladies sometimes condescend to employ for captivation. Whatever bears affinity to cunning is despicable" (PP, p. 44).

Women might well codify and advance their strategic knowledge in novels, but they are at a historical disadvantage: "Men have had every advantage of us in telling their own story. Education has been theirs in so much higher a degree; the pen has been in their hands" (P, p. 255). Indeed, the surest sign of Fanny's deprivation during her first weeks at Mansfield Park is that "she had not any paper" on which to write a letter to her brother (MP, p. 17). Given this disadvantage, women should remain in solidarity: "if the heroine of one novel be not patronized by the heroine of another, from whom can she expect protection and regard?... Let us not desert one another; we are an injured body" (NA, p. 30).

Thus to avoid charges of meanness and cunning, perhaps some misdirection is in order. Emma thinks that Mr. Elton's drunken proposal to her was a private mortification, but the seemingly foolish Miss Bates knows about it (and if Miss Bates knows, everyone knows; Vermeule [2010, p. 183] calls her "the Greek chorus of the novel, her words a vent through which its collective unconscious comes bubbling up"). When news is received that Mr. Elton has married, Miss Bates says, "Well, I had always rather fancied it would be some young lady hereabouts; not that I ever——Mrs. Cole once whispered to me—but I immediately said, 'No, Mr. Elton is a most worthy young man—but'——In short, I do not think I am particularly quick at those sort of discoveries. I do not pretend to it. What is before me, I see. At the same time, nobody could wonder if Mr. Elton should have aspired.——Miss Woodhouse lets me chatter on, so good-humouredly. She knows I would not offend for the world" (E, pp. 188–89). Here Miss Bates camouflages her insight with chatter and disclaimers. Catherine successfully hints to Henry and Eleanor Tilney that Isabella Thorpe has dumped her brother in favor of theirs. They sympathetically share her disappointment, and "Catherine, by some chance or other, found her spirits so very much relieved by this conversation, that she could not regret her being led on, though so unaccountably, to mention the circumstance which had produced it" (NA, p. 213). Catherine states for the record that it was not her doing; she was led on unaccountably. While exploring and theorizing strategic thinking in novels, perhaps women should best be not too explicit about it. As Elizabeth observes, "We all love to instruct, though we can teach only what is not worth knowing" (PP, p. 380).

CHAPTER TWELVE

Austen on Cluelessness

SOMETIMES PEOPLE do not understand that other people make their own decisions according to their own preferences. I call this "cluelessness," after the movie *Clueless* (1995), Amy Heckerling's adaptation of Austen's *Emma*. Austen's comprehensive analysis of strategic thinking extends to understanding its conspicuous absence.

Austen offers five explanations for cluelessness. The first is lack of natural ability: some people are not "naturally" geared towards strategic thinking and are inclined instead toward numbers, visual detail, literal meaning, and clear status distinctions. The second is social distance: an unmarried person for example is not so good at understanding married people because he has not yet had the experience of being married. Without lots of sincere communication, it is difficult to understand a person with quite different experiences. The third is excessive self-reference, using yourself too much as a template for understanding others. The fourth is status maintenance: a higher-status person is not supposed to think about the intentions of a lower-status person, and risks blurring the status distinction if she does. The fifth is that sometimes presumption, believing that one can directly manipulate another's preferences, actually works; you do not have to think about another person's preferences because you can change them. Finally, I apply these explanations to the decisive blunders of superiors such as Lady Catherine and General Tilney.

LACK OF NATURAL ABILITY

Austen emphasizes the importance of training for strategic thinking, but allows for personality differences. For example, Anne Elliot's strategic ability compared to Lady Russell's is due to a "natural" ability that trumps experience: "There is a quickness of perception in some, a nicety in the discernment of character, a natural penetration, in short, which no experience in others can equal, and Lady Russell had been less gifted in this part of understanding than her young friend" (P, p. 271).

As mentioned in chapter 2, being on the autistic spectrum is associated with weakness in understanding the mental states of others. Three other personality characteristics associated with autism are numeracy, attention to visual detail, and a fixation on literal meaning. Interestingly, we find these three characteristics in Austen's weak strategic thinkers.

The first characteristic is numeracy, exemplified by Mr. Collins and Mr. Rushworth. Mr. Collins specifies exact numbers when showing Elizabeth his garden: "[E]very view was pointed out with a minuteness which left beauty entirely behind. He could number the fields in every direction, and could tell how many trees there were in the most distant clump," and tries to impress Elizabeth with "his enumeration of the windows in front of the house" of his patron Lady Catherine (PP, pp. 177, 182). Mr. Rushworth defends the weightiness of his part in *Lovers' Vows* by declaring that "*The Count* has two and forty speeches...which is no trifle," and helps Edmund play Anhalt by counting his speeches too (MP, pp. 169, 186). As mentioned in chapter 11, after Frank Churchill's gallant rescue of Harriet Smith, Emma thinks, "Could a linguist, could a grammarian, could even a mathematician have seen what she did...without feeling that circumstances had been at work to make them peculiarly interesting to each other?" (E, p. 362). Linguists and grammarians are bad at strategic thinking but mathematicians are even worse.

The second characteristic is attention to visual detail. A good strategist should of course carefully observe others (especially their eyes), but for Austen, attention to visual detail, usually in clothing and physical appearance, is associated with weak strategic thinking. For Mrs. Allen, "[d]ress was her passion," and she spends days buying clothes to prepare for her and Catherine's first ball at Bath, only to have no idea what to do there other than wish that they knew someone (NA, p. 12). When Catherine eagerly apologizes to Henry Tilney for breaking her appointment with him and his sister, she appeals to Mrs. Allen for support: "You must have thought me so rude; but indeed it was not my own fault,—was it, Mrs. Allen? Did not they tell me that Mr. Tilney and his sister were gone out in a phaeton together? and then what could I do? But I had ten thousand times rather have been with you; now had not I, Mrs. Allen?" To help Catherine out, all Mrs. Allen has to do is agree, but she does not understand and instead replies, "My dear, you tumble my gown" (NA, pp. 92–93). Similarly, Mr. Rushworth is distracted easily by color and finery. His fiancée, Maria Bertram, shares several scenes with Henry Crawford in *Lovers' Vows* and prevents Mr. Rushworth from suspecting anything by "pointing out the necessity of his being very much dressed, and choosing his colours. Mr. Rushworth liked the idea of his finery very well, though affecting to despise it, and was too much engaged with what his own appearance would be, to think of the others" (MP, p. 162). Mr. Rushworth is excited: "I am to be Count Cassel, and am to come in first with a blue dress, and a pink satin cloak, and afterwards am to have another fine fancy suit by way of a shooting dress" (MP, p. 163). After Wickham has finally agreed to marry Lydia, Mrs. Bennet does not think about why Wickham made this decision; instead, Mrs. Bennet worries about Lydia's wedding clothes, and straight away "proceed[s] to all the particulars of calico, muslin, and cambric" (PP, p. 338).

Attention to visual detail can be more than just vanity; some people think about social interaction in general through physical appearance. Catherine asks Mrs. Allen if Henry Tilney's parents are also visiting in Bath, and Mrs. Allen does not at first remember: "Yes, I fancy they are, but I am not quite certain. Upon recollection, however, I have a notion they are both dead; at least the mother is; yes, I am sure Mrs. Tilney is dead, because Mrs. Hughes told me there was a very beautiful set of pearls that Mr. Drummond gave his daughter on her wedding-day and that Miss Tilney has got now, for they were put by for her when her mother died" (NA, p. 65). If not for the beautiful pearls, Mrs. Allen would not recall if Mrs. Tilney were alive or dead. For Lady Bertram, the knowledge of her son Tom's illness is nothing compared to actually seeing him: "The sufferings which Lady Bertram did not see, had little power over her fancy; and she wrote very comfortably about agitation and anxiety, and poor invalids, till Tom was actually conveyed to Mansfield, and her own eyes had beheld his altered appearance" (MP, p. 495). Mr. Rushworth is inspired to improve Sotherton after seeing how a friend's estate, Compton, has been renovated: "I wish you could see Compton....The approach *now* is one of the finest things in the country. You see the house in the most surprising manner. I declare when I got back to Sotherton yesterday, it looked like a prison—quite a dismal old prison" (MP, p. 62). When his guests visit Sotherton to suggest improvements, Mr. Rushworth "scarcely risked an original thought of his own beyond a wish that they had seen his friend Smith's place" (MP, pp. 113–14). Mr. Rushworth understands his friend's place and his own in completely visual terms, and cannot talk about them in any other way; he can only wish that others see what he sees. In Bath, Anne's sister Elizabeth Elliot at first refuses to acknowledge Captain Wentworth, but later includes him in a party because he looks good: "Captain Wentworth would move about well in her drawing-room" (P, pp. 245–46).

A compelling physical appearance can lead even strategically skilled people into error. After Wickham claims that Mr. Darcy mistreated him, Elizabeth Bennet says that Wickham has "truth in his looks," and as for Jane Bennet, "it was not in her nature to question the veracity of a young man of such amiable appearance as Wickham" (PP, pp. 96, 95). But once she reads Mr. Darcy's letter explaining his family's history with Wickham, Elizabeth weighs Wickham's appearance against actual evidence: "His countenance, voice, and manner, had established him at once in the possession of every virtue. . . . She could see him instantly before her, in every charm of air and address; but she could remember no more substantial good than the general approbation of the neighbourhood" (PP, p. 228). Realizing their error, Elizabeth and Jane recognize the power of appearances. Jane exclaims, "Poor Wickham; there is such an expression of goodness in his countenance! such an openness and gentleness in his manner!" and Elizabeth, comparing Wickham to Mr. Darcy, quips, "There certainly was some great mismanagement in the education of those two young men. One has got all the goodness, and the other all the appearance of it" (PP, p. 250). Similarly, after Willoughby's late-night confession to Elinor, "[s]he felt that his influence over her mind was heightened by circumstances which ought not in reason to have weight; by that person of uncommon attraction, that open, affectionate, and lively manner which it was no merit to possess" (SS, p. 377). Mr. Bennet married the foolish Mrs. Bennet because he was "captivated by youth and beauty, and that appearance of good humour, which youth and beauty generally give" (PP, p. 262).

The third characteristic of Austen's weak strategic thinkers is a fixation on literality, what words and symbols sound and look like and their formal meaning, as opposed to what they mean in the social context in which they are spoken or written. For example, Mr. Bennet jokes that Mary is learned because she copies passages word for word from books: "Do you consider the forms of introduction, and the stress that is laid on them, as nonsense? . . . What say you, Mary? for you are a young lady of deep reflection, I know, and read great books and make extracts" (PP, p. 7). Mary works hard on her music but performs with a "pedantic air and conceited manner" and cannot see how others see her, eagerly continuing to sing despite Elizabeth's "significant looks and silent entreaties" for her to stop (PP, pp. 27, 112). In comparison, since Jane Fairfax and Frank Churchill cannot speak openly to each other while secretly engaged, "Jane speaks, in effect, through the piano" (Wiltshire 1997, p. 71). Mary is "deep in the study of thorough bass and human nature" (PP, p. 67). Thorough bass is a kind of musical notation that uses numbers to specify chords above a bass line, as shown in figure 5 in chapter 14. Mary understands music as notes on a page, not as communication between performer and audience, and studies human nature in similar terms. After all, Mr. Bennet appeals to Mary as an authority on the rules and forms of introduction, what one learns from etiquette books, as opposed to actual social interaction. According to Nugent (2009, p. 91), Mary Bennet is "one of the earliest examples of a nerd in a famous work of literature." When the Bennets receive a letter from Mr. Collins introducing himself and expressing his hope to make amends for being the one to inherit their home, Elizabeth and Mr. Bennet puzzle over what exactly he means by this. Mary, however, observes that "[i]n point of composition . . . his letter does not seem defective. The idea of the olive branch perhaps is not wholly new, yet I think it is well expressed" (PP, p. 71). Mary is interested not in Mr. Collins's intentions but in his letter's composition and one stock phrase in particular. Mary likes Mr. Collins and "rated his abilities much higher than any of the others; there was a solidity in his reflections which often struck her," a solidity that is the opposite of Elizabeth's "quickness" (PP, pp. 139, 16).

Emma's first big mistake can be blamed on her excessive attention to literality. Emma thinks that Mr. Elton must love Harriet because of his enthusiasm about Emma's portrait of her; he calls it a "precious deposit" as he takes it to London to be framed (E, p. 51). But Mr. Elton prizes not the portrait's literal content but Emma's authorship.

Mr. Woodhouse, like his daughter, sees the portrait literally and worries that in the painting Harriet "seems to be sitting out of doors, with only a little shawl over her shoulders—and it makes one think she must catch cold" (E, p. 51). In contrast, Mr. Elton praises Emma's artistic skill in placing Harriet out of doors: "I regard it as a most happy thought, the placing of Miss Smith out of doors. . . . I never saw such a likeness" (E, p. 50). Similarly, when looking at a print of two men in a boat in a printshop window in Bath, Admiral Croft remarks to Anne, "What queer fellows your fine painters must be, to think that any body would venture their lives in such a shapeless old cockleshell as that," and jokes, "laughing heartily," "I wonder where that boat was built!" (P, pp. 183–84). Admiral Croft wonders about the painter's expertise, not the two men's health, and knows that the boat's seaworthiness is not a real issue but something to laugh about.

Emma encourages Harriet to copy riddles for her own collection, like Mary Bennet's collection of extracts, and asks Mr. Elton to contribute one. Mr. Elton delivers a charade to Emma and Harriet in which the last two lines are "Thy ready wit the word will soon supply, May its approval beam in that soft eye!" and Emma concludes that this is surely a declaration of love for Harriet (E, p. 76). Because of the charade's intimate intent, Harriet feels that she cannot possibly copy it in her book, but Emma says it is fine as long as she cuts off the last two lines which direct the riddle specifically toward her: "They are not at all the less written you know, because you divide them" (E, p. 82). Emma is completely willing to remove the literal words of the charade from the social context indicated by its last two lines and Mr. Elton's personal delivery of it. Emma then proceeds to read the decontextualized version aloud to her father.

Mrs. Allen likes to talk to herself. "[W]hile she sat at her work, if she lost her needle or broke her thread, if she heard a carriage in the street, or saw a speck upon her gown, she must observe it aloud, whether there were anyone at leisure to answer her or not"; Mrs. Allen's talking does not vary with the social context (NA, p. 57). When she talks with Mrs. Thorpe, "in what they called conversation... there was scarcely ever any exchange of opinion, and not often any resemblance of subject" (NA, p. 30). For Mrs. Allen, conversation is about saying words. In contrast, Miss Bates overcontextualizes, inscribing meaning in the smallest personal details of a message's reception: "If she is standing in a particular posture when she hears a piece of news, her posture becomes at once a part of the event which it is her duty to hand down to tradition" (Simpson 1870 [1968], p. 259).

Again, the three characteristics of Austen's weak strategic thinkers—numeracy, attention to visual detail, and a fixation on literality—are common among people on the autistic spectrum. Ferguson Bottomer (2007, p. 113) finds many other autistic spectrum character traits in *Pride and Prejudice*: for example, Mr. Darcy's dislike of dancing is shared by people on the autistic spectrum, who often find it difficult to coordinate their body movements, especially with others. The concept of "autism" was first characterized in the 1940s, and has been contested ever since (Silverman 2012); Austen observed the confluence of these three personality characteristics much earlier.

These three characteristics can be understood broadly as an inclination toward decontextualized literal meaning, as opposed to "illocutionary" meaning, which depends on how statements are made, who says them, in what social setting, and so forth (Austin 1975). Numbers, colors, muslins, pearls, as well as words as studied by linguists and grammarians, have "solidity" like Mr. Collins: thirty-nine is thirty-nine in almost all contexts.

This "solidity" can be desirable. Even though all meanings are contestable, it is helpful for certain terms to be uncontingent and fixed. For example, when a head waiter asks me how many people are in my party and I reply "two people," there might be some fuzziness about what a "person" means in terms of restaurant seating—my spouse and I might say that we are a party of two to get a table more quickly, and then once we are seated, pull up a chair for our four-year-old—but there is little contextual variation in, and little reason to be strategic about, what "two" means. When Henry Tilney has to run off to prepare a feast for his father's and Catherine's upcoming visit, Catherine does not understand why such effort is necessary, because General Tilney's words were, "Whatever you may happen to have in the house will be enough" (NA, p. 216). Catherine thinks to herself, "[W]hy he should say one thing so positively, and mean another all the while, was most unaccountable! How were people, at that rate, to be understood?" (NA, p. 218).

Not everything should be strategic; some communications should be literal.

In social life, social roles and status are one way to create literal meaning. A person becomes a police officer after years of training and a long selection process, but in a given moment it is much easier to think that a person is a police officer because she is wearing a uniform and badge. For those not good at strategic thinking, or for anyone in an unfamiliar or complicated social setting, social roles are quite helpful. Even Mr. Woodhouse can take another person's perspective with the help of social roles: when Jane Fairfax is set to leave to become a governess for Mrs. Smallridge, he tells Emma, "You know, my dear, she is going to be to this new lady what Miss Taylor was to us" (E, p. 421). Jane Fairfax's future social relations can be predicted because she will occupy the same role, governess, that Miss Taylor did. When Mr. Woodhouse says to Emma, "a bride, especially, is never to be neglected. . . . A bride, you know, my dear, is always the first in company," he again thinks in terms of social roles, because he is not good at strategic thinking: when Emma jokes that his policy encourages people to marry, Mr. Woodhouse "could not understand" (E, p. 302).

Similarly, when Marianne falls ill under her care, Mrs. Jennings takes Mrs. Dashwood's perspective by placing herself in the same social role, mother of a sick daughter: "when Mrs. Jennings considered that Marianne might probably be to *her* what [Mrs. Jennings's daughter] Charlotte was to herself, her sympathy in *her* sufferings was very sincere" (SS, p. 354). Knox-Shaw (2004, p. 146) writes that "Mrs. Jennings's grief is redoubled by the way she thinks herself into Mrs. Dashwood's position." People on the autistic spectrum who are not good at theory of mind tasks quite easily employ gender, race, and class stereotypes (Hirschfeld, Bartmess, White, and Frith 2007; White, Hill, Winston, and Frith 2006).

Thus a reliance on social roles and status is a fourth characteristic of Austen's weak strategic thinkers, along with numeracy, attention to visual detail, and a fixation on literality. Mr. Collins has all four. His numeracy has been mentioned. He demonstrates his commitment to literality by reading out loud three pages word for word out of Fordyce's Sermons; according to Morris (1987, p. 142), Mr. Collins is a study of "the deadly effect of literalism." He demonstrates his attention to social status and dress simultaneously by telling Elizabeth, "Lady Catherine will not think the worse of you for being simply dressed. She likes to have the distinction of rank preserved" (PP, p. 182). Knox-Shaw (2004, p. 103) notes that "Austen's great brilliance in conceiving Collins was to join two seemingly incompatible stock-in-trades—the toadying lackey and the presumptuous lover—and reveal their congruence." Indeed, status obsession and an inability to understand women as making their own choices go together.

Sir Walter, Anne Elliot's father, has three of the four characteristics (all except numeracy). Sir Walter, "at fifty-four, was still a very fine man. Few women could think more of their personal appearance than he did....He considered the blessing of beauty as inferior only to the blessing of a baronetcy" (P, p. 4). His attentions to social status and physical appearance are also combined in his objections to the navy: "First, as being the means of bringing persons of obscure birth into undue distinction, and raising men to honours which their fathers and grandfathers never dreamt of; and secondly, as it cuts up a man's youth and vigour most horribly; a sailor grows old sooner than any other man" (P, p. 21). Before meeting Admiral Croft, Sir Walter expects that "his face is about as orange as the cuffs and capes of my livery" (P, p. 24). Sir Walter understands Admiral Croft visually, in terms of color, in particular an orange which indicates family status (Todd and Blank [2006, p. 346] note that the colors of a livery are those of the family arms). Sir Walter obsessively reads his own family's entry in the Baronetage, the literal representation of his social status: "he could read his own history with an interest which never failed....Walter Elliot, born March 1, 1760, married, July 15, 1784" (P, p. 3). Sir Walter updates the entry with new information such as his daughter Mary's marriage, writing in the book by hand to give his own words the same solidity as the printer's. Sir Walter agrees to Anne's marriage to Captain Wentworth because of "his superiority of appearance...assisted by his well-sounding name," which balances Anne's "superiority of rank" (P, p. 271).

Elizabeth Elliot thinks like her father; when Anne warns her that Mrs. Clay might have designs on their father, Elizabeth's reply combines social status and physical appearance: "Mrs. Clay...never forgets who she is...I can assure you, that upon the subject of marriage...she reprobates all inequality of condition and rank more strongly than most people....That tooth of her's! and those freckles! Freckles do not disgust me so very much as they do him" (P, p. 37). Anne argues for the primacy of strategy and counters, "There is hardly any personal defect...which an agreeable manner might not gradually reconcile one to," but for Elizabeth, physical appearance trumps strategy: "[A]n agreeable manner may set off handsome features, but can never alter plain ones" (P, p. 38).

Mary Crawford, Lady Bertram, and Mr. Rushworth also combine these characteristics. When she first meets Fanny, Mary Crawford must ascertain her status and asks Edmund, "Pray, is she out, or is she not?— I am puzzled. She dined at the Parsonage, with the rest of you, which seemed like being *out*; and yet she says so little, that I can hardly suppose she *is*." Edmund does not share Mary's fixation, replying, "I believe I know what you mean—but I will not undertake to answer the question. My cousin is grown up. She has the age and sense of a woman, but the outs and not outs are beyond me" (MP, p. 56). Mary is also concerned with the sound (not just the meaning) of Edmund's name when combined with a title: "There is something in the sound of Mr. *Edmund* Bertram so formal, so pitiful, so younger-brother-like, that I detest it.... *Lord* Edmund or *Sir* Edmund sound delightfully" (MP, p. 246). In a letter to Fanny, Mary Crawford brags about Edmund's looks ("my friends here are very much struck with his gentleman-like appearance"), and a chagrined Fanny thinks to herself, "The woman who could speak of him, and speak only of his appearance!—What an unworthy attachment!...*She* who had known him intimately half a year!" (MP, pp. 482, 484). Lady Bertram understands Henry Crawford's proposal to Fanny in terms of beauty and status: "To know Fanny to be sought in marriage by a man of fortune, raised her, therefore, very much in her opinion. By convincing her that Fanny *was* very pretty...it made her feel a sort of credit in calling her niece....And looking at her complacently, she added, 'Humph—We certainly are a handsome family'" (MP, pp. 383–84). Lady Bertram reduces the entire social process that led to Henry's proposal to Fanny's physical appearance, which enhances her own status. Mr. Rushworth equates status with appearance, which is in turn reducible to a precise numerical height: when Tom Bertram tells his father that Henry Crawford is a "gentleman-like man," Mr. Rushworth interjects, "I do not say he is not gentleman-like, considering; but you should tell your father he is not above five feet eight, or he will be expecting a well-looking man" (MP, p. 217).

When Mr. Bingley suggests that Mr. Darcy ask Elizabeth to dance, Mr. Darcy replies, "She is tolerable; but not handsome enough to tempt *me*; and I am in no humour at present to give consequence to young ladies who are slighted by other men" (PP, p. 12). Elizabeth hears this remark, and thus begins their mutual dislike. Mr. Darcy normally has good strategic skills but has just arrived in the neighborhood. In this completely new social environment, Elizabeth's appearance and her status as unwanted are all he has to go on, and this might explain why "his extreme insolence, at the first Meryton ball, does not quite match his later behavior" (Kennedy 1950, p. 53, quoted in Morris 1987, p. 159).

Perhaps being male is a fifth related characteristic of weak strategic thinkers: as mentioned in chapter 2, some interpret autism as an "extreme male brain." It is true that Austen is more willing to accept cluelessness in a grown man than in a grown woman: for example, Mr. Bingley, "so easily guided that his worth was invaluable," is subject to a milder degree of ridicule than a woman who is equally weak strategically, such as Mary Musgrove (PP, p. 412). There is no female equivalent of Mr. Woodhouse, a person whose cluelessness is annoying but sometimes endearing; Lady Russell's lack of strategic ability is just a lack, with no upside or charm, and if Mrs. Dashwood is considered endearing, it is because of her overstrategicness, not her cluelessness (although her overstrategicness might be considered a higher level of cluelessness). It is also true that Henry Tilney, whose theoretical statements about strategic thinking are most explicit, understands muslins: "[M]y sister has often trusted me in the choice of a gown. I bought one for her the other day, and it was pronounced to be a prodigious bargain by every lady who saw it. I gave but five shillings a yard for it, and a true Indian muslin." If there is any doubt that this skill is gendered, Mrs. Allen responds: "Men commonly take so little notice of those things.... You must be a great comfort to your sister, sir" (NA, p. 20). Catherine laughs and almost calls him strange; according to Woods (1999, p. 138), "subliminally, Jane Austen has conjured up the spectacle of Henry Tilney *in drag.*"

Still, for Austen the gendering of cluelessness is anything but simple. Being strategically skilled does not necessarily make you less male. Mr. Knightley understands the needs of Jane Fairfax and Miss Bates and arranges for his carriage to take them home after a party. Mrs. Weston exclaims, "Such a very kind attention—and so thoughtful an attention!— the sort of thing that so few men would think of" (PP, p. 241). Mr. Knightley's thoughtfulness might be unusual for a man, but in this case, his ability to anticipate and provide for women's needs if anything makes him more, not less, of a man. The only difference seems to be that Henry Tilney anticipates clothing needs while Mr. Knightley anticipates transportation needs.

Perhaps strategic thinking is gendered according to the arena in which it is employed. Admiral Croft might be in charge at sea, but "on land, the Admiral defers to Sophia's [Mrs. Croft's] greater expertise" (Mellor 2000, pp. 130–31). Similarly, Captain Wentworth captures French frigates and succeeds brilliantly as a naval officer, but is much less brilliant in the arena of courtship, in which "[h]is hope of domestic happiness has been thwarted by his excessive pride" (Mellor 2000, p. 125). Captain Wentworth calls out to Anne for help when Louisa Musgrove falls unconscious because "[t]his is not the kind of action he has been trained for" (Sutherland and Le Faye 2005, p. 224). Luckily, on land Anne's strategic skills pull them through.

As to whether women or men are better at strategic thinking in general, when Catherine Morland and Eleanor Tilney misunderstand each other (Catherine says "something very shocking indeed, will soon come out in London," but she means a novel, not an actual riot), Henry Tilney jokingly concedes that women have a natural advantage: "[N]o one can think more highly of the understanding of women than I do. In my opinion, nature has given them so much, that they never find it necessary to use more than half" (NA, pp. 113, 115). If a woman has greater strategic capacity, a man might more than compensate through greater effort.

SOCIAL DISTANCE

Sex differences can cause cluelessness because it is not easy to communicate across them, not because one sex is necessarily more skilled than another. Had she more time, Catherine would have bought a new gown for the cotillion ball to catch the eye of Henry Tilney, but "[t]his would have been an error in judgment. . . . It would be mortifying to the feelings of many ladies, could they be made to understand how little the heart of man is affected by what is costly or new in their attire." A woman learns this mortifying fact only by getting to know men: "a brother rather than a great aunt, might have warned her, for man only can be aware of the insensibility of man towards a new gown" (NA, p. 71). After all, Henry Tilney knows about muslins because he has a sister. Actually, perhaps buying a new gown would not have been an error, given Henry's eye for fashion.

Austen shows how communicating across sex differences takes real effort. When Anne and Captain Harville argue over which sex loves with greater constancy, Captain Harville first appeals to analogies ("as our bodies are the strongest, so are our feelings; capable of bearing most rough usage, and riding out the heaviest weather") and next to literature ("Songs and proverbs, all talk of woman's fickleness"), but finally breaks through by communicating directly, like a brother, how a man feels: "[I]f I could but make you comprehend what a man suffers when he takes a last look at his wife and children, and watches the boat that he has sent them off in, as long as it is in sight, and then turns away and says, 'God knows whether we ever meet again!' And then, if I could convey to you the glow of his soul when he does see them again. . . . If I could explain to you all this, and all that a man can bear and do, and glories to do for the sake of these treasures of his existence!" (P, pp. 253–55). Feeling the warmth of Captain Harville and men in general, Anne concedes, "God forbid that I should undervalue the warm and faithful feelings of any of my fellow-creatures." Anne is more than willing to take a man's point of view, acknowledging, "You are always labouring and toiling, exposed to every risk and hardship. . . . It would be too hard indeed . . . if woman's feelings were to be added to all this" (P, pp. 256, 254). With enough effort and sincerity, women and men can understand each other, and

Anne's willingness to do this is rewarded by the love of the eavesdropping Captain Wentworth.

In general, social distance decreases communication and increases cluelessness. Along with sex differences, Austen also considers differences in location, marital status, and age. When her family leaves Kellynch Hall, Anne goes to stay with her sister Mary at Uppercross Cottage, a different physical location. Anne well knows from previous visits that "a removal from one set of people to another, though at a distance of only three miles, will often include a total change of conversation, opinion, and idea" and that no one at Uppercross thinks about "the affairs which at Kellynch Hall were treated as of such general publicity and pervading interest" (P, p. 45). Noticing the lack of sympathy at Uppercross for her family's decampment to Bath, Anne learns "another lesson, in the art of knowing our own nothingness beyond our own circle.... She could only resolve to avoid such self-delusion in future" (P, pp. 45–46).

Without listening to each other, even people in the same house can be unaware of how others see them. When the company at Mansfield Park rehearses *Lovers' Vows*, Fanny "knew that Mr. Yates was in general thought to rant dreadfully, that Mr. Yates was disappointed in Henry Crawford, that Tom Bertram spoke so quick he would be unintelligible, that Mrs. Grant spoiled everything by laughing, that Edmund was behind-hand with his part, and that it was misery to have any thing to do with Mr. Rushworth.... [S]he found every body requiring something they had not, and giving occasion of discontent to the others.... [N]obody but the complainer would observe any directions" (MP, p. 193).

Captain Wentworth tells his sister Mrs. Croft that he does not like women and children aboard his ship because of the difficulty of giving them properly comfortable accommodations. Mrs. Croft replies, "We none of us expect to be in smooth water all our days," and Admiral Croft agrees that "when he had got a wife, he will sing a different tune. When he is married...we shall see him do as you and I, and a great many others, have done. We shall have him very thankful to anybody that will bring him his wife" (P, p. 75). The Crofts claim that as an unmarried man, Captain Wentworth simply does not understand married people and even what he himself will be like when married. Captain Wentworth expresses his frustration with communicating over this divide: "When once married people begin to attack me with, 'Oh! you will think very differently, when you are married.' I can only say, 'No, I shall not;' and then they say again, 'Yes, you will,' and there is an end of it" (P, p. 76).

Finally, Mr. Woodhouse observes that Mr. John Knightley, brother of Mr. Knightley, is too rough with his children, and Emma explains, "He appears rough to you...because you are so very gentle yourself.... The children are all fond of him." Mr. Woodhouse cannot understand the children's enjoyment when Mr. Knightley "tosses them up to the ceiling in a very frightful way!" (E, p. 86). Emma replies, "That is the case with us all, papa. One half of the world cannot understand the pleasures of the other" (E, p. 87). Mr. Woodhouse's understanding cannot overcome the difference in age (and disposition) between himself and the children; as Emma observes, the problem of understanding other people's preferences is universal, not just between young and old.

At least Mr. Woodhouse has Emma as an interlocutor, who talks "slowly and distinctly, and two or three times over, with explanations of every part," just as Captain Wentworth has Captain Harville (E, p. 83). Similarly, Sir Thomas, when trying to persuade Fanny to accept Henry Crawford's proposal, "did not understand her; he felt that he did not; and therefore applied to Edmund to tell him how she stood affected on the present occasion" (MP, p. 422). In Native American tribes, "two spirit" people traditionally serve as community mediators, "stand[ing] between the men and the women because we know what it is to be both" (Gilley 2006, p. 169; see also Evans-Campbell, Fredriksen-Goldsen, Walters, and Stately 2007, and Roth 2012).

EXCESSIVE SELF-REFERENCE

Emma points out that Mr. Woodhouse is clueless because he thinks of others' preferences only in terms of his own; he cannot understand the joys of rough play because of his own gentleness. Mr. Woodhouse is "never able to suppose that other people could feel differently from himself" and therefore believes, for example, that Emma's governess Miss Taylor would rather stay with them than start her new married life as Mrs. Weston (E, p. 6). At the wedding, "he could never believe other people to be different from himself. What was unwholesome to him he regarded as unfit for any body; and he had, therefore, earnestly tried to dissuade them from having any wedding-cake at all" (E, p. 17). Austen sets up Mr. Woodhouse to fail the false belief test (described in chapter 2). Similarly, "Mrs. Allen, not being at all in the habit of conveying any expression herself by a look, was not aware of its being ever intended by any body else" (NA, p. 57). We also have "Lady Bertram holding exercise to be as unnecessary for every body as it was unpleasant to herself; and Mrs. Norris, who was walking all day, thinking every body ought to walk as much" (MP, p. 41).

Focusing too much on yourself hurts your strategic thinking. Bhatt and Camerer (2005, p. 446) look at the brain activity of a person playing a game with another and find that good performers "are better at imagining what others will do, and this imaginative process...uses all-purpose circuitry that is generally used in creating empathy or doing emotional forecasting involving others," while "subjects who are more self-focussed do not think enough about the other player and make poorer choices and less accurate guesses." For another example, the nuclear strategist Albert Wohlstetter, senior analyst at the RAND Corporation, "assumed that people on the other side thought like RAND analysts; it was the mirror-imaging problem of intelligence agencies, the presumption that when the other side weighs the risks and possible benefits, it comes to the same rational conclusion as your side does" (Abella 2008, p. 121). Similarly, Poe's (1845 [1998], p. 258) policemen "consider only their *own* ideas of ingenuity; and, in searching for anything hidden, advert only to the modes in which *they* would have hidden it." To some extent this is entirely natural: Ames, Jenkins, Banaji, and Mitchell (2008) find that when a person is asked to put himself in the place of another, the regions of his brain that are active are similar to the regions that are active when he is asked introspective questions. However, according to Nickerson (1999, p. 749), "people's tendency to see their own knowledge, beliefs, attitudes, and actions as more representative of those of others than they really are, are examples of a useful heuristic being applied in a less than optimal way."

Austen illustrates how cluelessness can result from using a single person, yourself, to estimate the entire range of human variety. For people as mature as Mr. Woodhouse and Mrs. Allen, this ignorance can be ascribed to personality traits or incapacity. For people as young as Fanny Price and Catherine Morland, it can result from inexperience. When Fanny refuses Henry Crawford's proposal, "Fanny knew her own meaning, but was no judge of her own manner. Her manner was incurably gentle, and she was not aware how much it concealed the sternness of her purpose" (MP, pp. 377–78). Fanny does not know how wimpy she sounds to others, as she has never had to refuse anyone before. As mentioned in chapter 5, when Catherine tells Henry Tilney that his brother Captain Tilney must have wanted to dance with the engaged Isabella because of his friendly good nature, Henry replies that Catherine is thinking about what she herself would do, not what another quite different person would do. To Henry this shows Catherine's good character: "[Y]our attributing my brother's wish of dancing with Miss Thorpe to good-nature alone, convinced me of your being superior in good-nature yourself" (NA, p. 135).

Excessive self-reference is not always charming. By referring excessively to yourself, you can efface the preferences of others. For example, when invited to the Coles' party, Mr. Woodhouse subsumes Emma's wishes into his own: "I am not fond of dinner-visiting...I never was. No

more is Emma. Late hours do not agree with us" (E, p. 225). Elinor scolds
Edward Ferrars for visiting and giving her and her family the impression
that he was seeking her affection, even while he was secretly engaged to
Lucy Steele. But Edward thinks only of his own conviction, not about
how his visit was perceived by others: "I was simple enough to think,
that because my *faith* was plighted to another, there could be no danger
in my being with you. . . . [T]he arguments with which I reconciled myself
to the expediency of it, were no better than these:—The danger is my
own; I am doing no injury to anybody but myself" (SS, p. 417). Edward,
thinking that his actions injure only himself, gave Elinor real pain. Sir
Thomas is surprised to find that Fanny has never had a fire in her room,
even in winter, but defends Mrs. Norris's decision: "Your aunt Norris
has always been an advocate, and very judiciously, for young people's
being brought up without unnecessary indulgences. . . . She is also very
hardy herself, which of course will influence her in her opinion of the
wants of others" (MP, p. 361). According to Sir Thomas, Mrs. Norris
either did not believe that Fanny should have the indulgence of a fire or,
because of excessive self-reference, did not consider the possibility that
Fanny might want one. Fanny freezes regardless.

HIGH-STATUS PEOPLE ARE NOT SUPPOSED TO ENTER THE MINDS OF LOW-STATUS PEOPLE

Actually, Sir Thomas's two reasons why Fanny does not have a fire are
not always distinct: sometimes you have difficulty understanding how
another person thinks because you believe your own way of thinking
is better. Austen makes this point in several examples. When Mary
Musgrove learns that Captain Benwick has proposed to Louisa, she
writes in a letter to Anne, "Are not you astonished? I shall be surprised at
least if you ever received a hint of it, for I never did" (P, p. 179). Because
she considers herself better at detecting than her sister Anne, Mary does
not think it possible that Anne could have detected it first. Mrs. Elton
obnoxiously compares everything in Highbury to her brother-in-law's
residence, Maple Grove ("That room was the very shape and size of
the morning-room at Maple Grove"), and by sheer repetition forces it
into the discourse of Highbury residents, including at first Miss Bates
and eventually even Emma (E, p. 294). It is possible that Mrs. Elton is
simply incapable of thinking about anything except in reference to Maple
Grove, but more likely she is attempting to get Highbury residents to take
for granted its superiority. When Elizabeth cannot believe that Charlotte
Lucas has accepted Mr. Collins's proposal, Charlotte responds, "Why
should you be surprised, my dear Eliza?—Do you think it incredible

that Mr. Collins should be able to procure any woman's good opinion, because he was not so happy as to succeed with you?" (PP, p. 140). Elizabeth does not restrain herself from thinking that Charlotte made a terrible decision: "Charlotte the wife of Mr. Collins, was a most humiliating picture!—And to the pang of a friend disgracing herself and sunk in her esteem, was added the distressing conviction that it was impossible for that friend to be tolerably happy in the lot she had chosen" (PP, p. 141). After Harriet finally agrees to marry Robert Martin, "Emma could now acknowledge, that Harriet had always liked Robert Martin; and that his continuing to love her had been irresistible.—Beyond this, it must ever be unintelligible to Emma" (E, p. 525). By thinking Harriet's love unintelligible, Emma still maintains that Harriet is not making the best choice. Compare this with Charles Musgrove's generous spirit when telling Anne, "I hope you do not think I am so illiberal as to want every man to have the same objects and pleasures as myself" (P, p. 237).

Austen illustrates how not thinking it necessary to understand another person's thought processes can be a mark of superior status over that person. Elizabeth's unwillingness to understand Charlotte's decision to marry Mr. Collins is more than just an expression that Charlotte is obviously mistaken; Elizabeth also thinks that Charlotte is "disgracing herself and sunk in her esteem." Similarly, in deeming Harriet's desire unintelligible, Emma gives up her pet project of elevating Harriet's social status and is content that it remain where it is. When the Dowager Viscountess Dalrymple arrives in Bath, Anne is dismayed to see how anxious her father Sir Walter and sister Elizabeth are to introduce themselves: "I confess it does vex me, that we should be so solicitous to have the relationship acknowledged, which we may be very sure is a matter of perfect indifference to them" (P, p. 163). Similarly, when Admiral Croft arrives in Bath, Sir Walter "think[s] and talk[s] a great deal more about the Admiral, than the Admiral ever thought or talked about him" (P, p. 182). When Mr. Elton proposes, what really infuriates Emma is his presumption that she is thinking of him: "that he should talk of encouragement, should consider her as aware of his views, accepting his attentions, meaning (in short), to marry him! . . . It was most provoking." Emma regards herself as a person of higher status, who of course does not think about Mr. Elton's thought processes. Mr. Elton is violating rank: "he must know that in fortune and consequence she was greatly his superior. He must know that the Woodhouses had been settled for several generations at Hartfield . . . and that the Eltons were nobody" (E, p. 147). It is outrageous for Mr. Elton to presume that Emma is thinking of him, but Emma can freely specify what Mr. Elton should know.

When Sir Walter has to leave Kellynch Hall to reduce expenses, he "could not have borne the degradation of being known to design letting his house. . . . Sir Walter spurned the idea of its being offered in any manner; forbad the slightest hint being dropped of his having such an intention; and it was only on the supposition of his being spontaneously solicited by some most unexceptionable applicant, on his own terms, and as a great favor, that he would let it at all" (P, pp. 16–17). Advertising Kellynch Hall would show to the world that Sir Walter is thinking about whether other people are interested, which is a degradation; if a prospective tenant spontaneously solicits, however, then the tenant must think about whether Sir Walter will accept.

Since superiors should not think of the minds of inferiors, they need agents to do their dirty work for them, such as the lawyer Mr. Shepherd, who finds a tenant to spontaneously solicit Sir Walter. When Mrs. Rushworth invites Lady Bertram to join in the visit to Sotherton, "Lady Bertram constantly declined it; but her placid manner of refusal made Mrs. Rushworth still think she wished to come, till Mrs. Norris's more numerous words and louder tone convinced her of the truth" (MP, p. 89). Like Fanny, Lady Bertram does not know how feeble she sounds when refusing an offer, but unlike Fanny, she never has to stoop to think about whether others understand her, because she has Mrs. Norris to be loud on her behalf.

Such an agent can be imaginary. Sir Thomas cannot help but notice Henry Crawford's interest in Fanny, even though he is above it all: "though infinitely above scheming or contriving for . . . the most advantageous matrimonial establishment . . . he could not avoid perceiving in a grand and careless way that Mr. Crawford was somewhat distinguishing his niece—nor perhaps refrain (though unconsciously) from giving a more willing assent to invitations on that account. . . . [A]ny one in the habit of such idle observations *would have thought* that Mr. Crawford was the admirer of Fanny Price" (MP, pp. 277–78). Sir Thomas's altitude is preserved because he acts carelessly and even unconsciously; he is not putting any effort into noticing, because any imagined person would idly notice.

The idea that superiors do not think about what inferiors are thinking about is even applied by Austen to sexes, animals, and inanimate objects. "[I]t must be very improper that a young lady should dream of a gentleman before the gentleman is first known to have dreamt of her" (NA, p. 22). Similarly, among women, "It is a thing of course among us, that every man is refused—till he offers"; a woman should not have to think about a man's decision-making process (P, p. 212). Instead of understanding his horse, John Thorpe tells Catherine that "he will soon know his master" (NA, p. 58). When Marianne leaves the Norland estate, she bids it goodbye: "No leaf will decay because we are removed, nor any branch become motionless although we can observe you no longer!—No; you will continue the same; unconscious of the pleasure or the regret you occasion, and insensible of any change in those who walk under your shade!" (SS, p. 32).

PRESUMPTION SOMETIMES WORKS

Austen's fifth explanation for cluelessness is simply that if you can change another person's preferences at your will, you don't have to care about them. *Pride and Prejudice* famously begins with a strong presumption: "It is a truth universally acknowledged, that a single man in possession of a good fortune, must be in want of a wife. However little known the feelings or views of such a man may be on his first entering a neighbourhood, this truth is so well fixed in the minds of the surrounding families, that he is considered the rightful property of some one or other of their daughters" (PP, p. 3). Does this mean that neighborhood families are clueless? Are they wrong to think that the feelings and views of their newly arrived single men are irrelevant? Sometimes a young woman can be delusional, as when Elizabeth thinks that she was "the first to excite and to deserve [Wickham's] attention" (PP, p. 172). But sometimes a young woman presumes correctly, as when Elizabeth's "fancy told her she still possessed" the "power...of bringing on the renewal of [Mr. Darcy's] addresses" (PP, p. 293). When Mrs. Gardiner warns Elizabeth not to get involved with Wickham because he has no fortune, Elizabeth agrees, "I will not be in a hurry to believe myself his first object. When I am in company with him, I will not be wishing" (PP, p. 164). Wickham's own beliefs and preferences are irrelevant, because presumably he can be led on entirely by Elizabeth's wishing. Indeed, recall that Catherine creates Henry Tilney's attachment outright: "a persuasion of her partiality for him had been the only cause of giving her a serious thought" (NA, p. 252). Sometimes presumption works.

DECISIVE BLUNDERS

Austen is interested in cluelessness not just as a phenomenon in itself, but also as the cause of decisive and life changing blunders, by superiors especially. When Lady Catherine meets with Elizabeth, she immediately puts her on notice: "You can be at no loss, Miss Bennet, to understand the reason of my journey hither. Your own heart, your own conscience, must tell you why I come" (PP, p. 391). By asserting that Elizabeth must know what she is thinking, Lady Catherine tries to establish Elizabeth's inferior status. Lady Catherine has heard a report that her nephew

Mr. Darcy and Elizabeth might be married, and wants Elizabeth to contradict it. Elizabeth replies with strategic reasoning: "Your coming to Longbourn, to see me and my family . . . will be rather a confirmation of it; if, indeed, such a report is in existence" (PP, p. 392). Lady Catherine argues in terms of status: she and Mr. Darcy's mother had planned for him to marry her daughter, his cousin Miss de Bourgh, and in this sense they have long had the status of being engaged (regardless of the actual preferences of Mr. Darcy or Miss de Bourgh). Lady Catherine is fixated on literality: she wants Elizabeth to directly answer whether she is engaged to Mr. Darcy and explicitly promise that she will never enter into such an engagement. Elizabeth finally answers that she is not, but refuses to make such a promise, again arguing strategically, "Supposing him to be attached to me, would *my* refusing to accept his hand, make him wish to bestow it on his cousin?" (PP, p. 395). Lady Catherine mentions the "patched-up business" of Lydia's marriage to Wickham, and resorts to caste language: "Are the shades of Pemberley to be thus polluted?" (PP, p. 396).

Lady Catherine later tells Mr. Darcy of Elizabeth's rude refusal, to impress upon him Elizabeth's impudence and disregard of social rank, "dwelling emphatically on every expression . . . which, in her lady-ship's apprehension, peculiarly denoted her perverseness and assurance" (PP, p. 407). Mr. Darcy thereby learns that Elizabeth has not completely written him off, which gives him the confidence to renew his proposal: "My aunt's intelligence had given me hope" (PP, p. 423). Lady Catherine thinks that she told Mr. Darcy the literal fact of Elizabeth's refusal to promise not to marry him, but the true meaning of Elizabeth's refusal is understandable only in the full context of Elizabeth and Mr. Darcy's shared history, which Mr. Darcy understands: "I knew enough of your disposition to be certain, that, had you been absolutely, irrevocably decided against me, you would have acknowledged it to Lady Catherine, frankly and openly." Elizabeth understands that Mr. Darcy understands: "Yes, you know enough of my *frankness* to believe me capable of *that*. After abusing you so abominably to your face, I could have no scruple in abusing you to all your relations" (PP, p. 407).

If Lady Catherine had thought Elizabeth in the least bit strategic, she would not have told Mr. Darcy that Elizabeth refused to promise not to marry him, especially given that she already knew that there was sufficient sentiment on his side to make their marriage plausible. Invested in her own status and assuming that her nephew Mr. Darcy is just like her, Lady Catherine blunders decisively.

When General Tilney finds that Catherine Moreland is not an heiress, he responds to this status violation by ritually expelling her, throwing her out of Northanger with no notice and no escort home. He could have ended Catherine's visit more quietly and gently, but instead he wanted to make a performative statement: "Enraged with almost every body in the world but himself, he set out the next day for the Abbey, where his performances have been seen" (NA, p. 256). But of course this is another decisive blunder: "Henry's indignation on hearing how Catherine had been treated...had been open and bold....He felt himself bound as much in honour as in affection to Miss Morland, and believing that heart to be his own which he had been directed to gain, no unworthy retraction of a tacit consent, no reversing decree of unjustifiable anger, could shake his fidelity, or influence the resolutions it prompted" (NA, p. 257).

General Tilney knew of Henry's affection toward Catherine, having promoted it himself, and could have anticipated that such callous mistreatment would only increase Henry's sympathy by adding feelings of injustice. "The General, accustomed on every ordinary occasion to give the law in his family, prepared for no reluctance but of feeling, no opposing desire that should dare to clothe itself in words," does not think of Henry as a person who has his own desires and makes his own decisions; it does not occur to him that Henry might actually do something and thus he is completely unprepared for Henry's opposition (NA, p. 257). General Tilney thinks that Henry's mental state is something that he can control: he "ordered [him] to think of her no more" (NA, p. 253). For that matter, General Tilney's original mistake of thinking Catherine an heiress is also due to his inability to understand the objectives of his informant John Thorpe: "pretty well resolved upon marrying Catherine himself," John Thorpe's "vanity induced him to represent the family as yet more wealthy than his vanity and avarice had made him believe them" (NA, p. 254). Eager to hear John Thorpe's literal statement that Catherine is the heir of the Allens, General Tilney does not consider his motives for saying so. Finally, General Tilney's sending Catherine home without a servant to accompany her (and report back) provides Henry with the perfect excuse to go and visit her, to make sure that she made it home safely.

Mrs. Ferrars, Edward's mother, responds to his secret engagement with Lucy Steele by slashing his status, replacing him as favored eldest son with his brother Robert only to have Lucy marry Robert instead. As Elinor points out to Edward, "[Y]our mother has brought on herself a most appropriate punishment. The independence she settled on Robert, through resentment against you, has put it in his power to make his own choice; and she has actually been bribing one son with a thousand a-year, to do the very deed which she disinherited the other for intending to do" (SS, p. 414). This is another case of a privileged, status-obsessed person whose action backfires because he or she does not take into account the strategicness of a person of lower status, in this case Lucy. To be fair, no one, including Elinor and Marianne, anticipated that Lucy would capture Robert, and so Mrs. Ferrars does not blunder as decisively as do Lady Catherine and General Tilney.

Our three blundering elder superiors share some of the other characteristics of cluelessness. Lady Catherine knows nothing of music and speaks about it in terms of status: "Our instrument is a capital one, probably superior to—.... There are few people in England, I suppose, who have more true enjoyment of music than myself, or a better natural taste. If I had ever learnt, I should have been a great proficient" (PP, pp. 185, 194). Lady Catherine's insistence to know Elizabeth's exact age ("You cannot be more than twenty, I am sure") possibly indicates her numeracy (PP, p. 187). When Catherine Morland arrives at Northanger Abbey, General Tilney asks whether she would like to see the house or the grounds first, and even though Catherine is anxious to see the house, "he certainly read in Miss Morland's eyes a judicious desire of making use of the present smiling weather" (NA, p. 181). When he later gives Catherine a tour of the house, he "satisfied his own curiosity, in a close examination of every well-known ornament" and "could not forego the pleasure of pacing out the length" of the dining room (NA, pp. 187–188). General Tilney reads eyes poorly, likes visual detail, and is numerate. We know less about Mrs. Ferrars, although she "particularly requested to look at" a pair of screens before being informed that they were painted by Elinor (SS, p. 268). Since Mrs. Ferrars rarely expresses interest in anything, perhaps this shows an attention to visual detail.

Our next two blunders are by young people who think themselves superior. Henry Crawford works his way up to his decisive blunder. Henry declares that he knows Fanny's mind: "I do and will deserve you; and when once convinced that my attachment is what I declare it, I know you too well not to entertain the warmest hopes" (MP, p. 398). Henry even places words in Fanny's mouth: "I fancied you might be going to tell me I *ought* to be more attentive" (MP, p. 394). But he does not have a clue about how much Fanny hates him. After trying to convince Fanny, Henry says that he is reluctant to leave, "but there was no look of despair in parting to bely his words," and his presumption makes Fanny "angry. Some resentment did arise at a perseverance so selfish and ungenerous" (MP, p. 379). Later Henry says he is interested in the welfare of his poorer tenants, which "was aimed, and well aimed, at Fanny. It was pleasing to hear him speak so properly.... [S]he was on the point of giving him an approving look when it was all frightened off, by his adding a something too pointed of his hoping soon to have an assistant, a friend, a guide in every plan of utility or charity" (MP, pp. 469–70). Certain that he will succeed, Henry does not know when to shut up. Finally and decisively, Henry does not realize the extent of Maria's feelings when he visits her after her marriage and ends up running off with her, "entangled by his own vanity" (MP, p. 541).

Emma regards Harriet Smith as transparent and moldable, and only when Harriet suggests that Mr. Knightley might be interested in her does Emma realize that Harriet is capable of independent thought. Emma realizes how badly her plan backfires: "Oh! had she never brought Harriet forward!... Had she not, with a folly which no tongue could express, prevented her marrying the unexceptionable young man who would have made her happy and respectable in the line of life to which she ought to belong—all would have been safe" (E, p. 450). Even though she herself had encouraged Harriet to marry into higher status, Emma is shocked at Harriet's not knowing her own place: "How Harriet could ever have had the presumption to raise her thoughts to Mr. Knightley!— How she could dare to fancy herself the chosen of such a man till actually assured of it!" (E, pp. 450–51).

We have seen Emma's orientation toward literality in her willingness to separate and decontextualize the lines of Mr. Elton's riddle and in her literal, not socially contextualized, understanding of Mr. Elton's admiration of her portrait of Harriet. Henry Crawford offers to Edmund Bertram his advice about how to improve Thornton Lacey, where Edmund will live as a clergyman, specifically recommending that the farmyard be eliminated and the blacksmith's shop visually obscured; as Knox-Shaw (2004, p. 94) observes, these alterations " 'shut out' all traces of the vicarage's social context."

The closest thing to a decisive blunder in *Persuasion* is Lady Russell's attempt to persuade Anne to succeed her mother as Lady Elliot by marrying Mr. Elliot, which would have been disastrous. Lady Russell tells Anne, "You are your mother's self in countenance and disposition; and if I might be allowed to fancy you such as she was, in situation, and name, and home, presiding and blessing in the same spot, and only superior to her in being more highly valued!" (P, p. 173). Lady Russell's scenario is both visually detailed and status-oriented. Lady Russell has "prejudices on the side of ancestry; she had a value for rank and consequence, which blinded her a little to the faults of those who possessed them" (P, p. 12). After Anne marries Captain Wentworth, Lady Russell realizes that "she had been unfairly influenced by appearances... that because Captain Wentworth's manners had not suited her own ideas, she had been too quick in suspecting them" (P, p. 271). Lady Russell had indulged in visuality and excessive self-reference.

Lady Russell cannot take all the blame for their eight-year delay, however. As mentioned in chapter 9, when Captain Wentworth asks Anne if she would have accepted him had he proposed again in two years, after he had achieved some success in the navy, Anne replies that of course she would have. Captain Wentworth laments, "It is not that I did not think of it, or desire it, as what could alone crown all my other success. But I was proud, too proud to ask again. I did not understand you. I shut my eyes, and would not understand you" (P, p. 268). Captain Wentworth actively considered proposing again, but pride prevents him from thinking about what Anne would do. Considering Anne a crown for his other successes also shows some concern for status. But the main cause of Captain Wentworth's cluelessness about Anne is the social distance between them, between sea and land, between man and woman, which they gradually bridge, with the help of their mutual friends.

CHAPTER THIRTEEN

Real-World Cluelessness

BUILDING UPON Austen's analysis, I find real-world examples of clue-lessness and come up with five explanations of my own. First, cluelessness can be considered as just another kind of mental laziness; still, I argue that cluelessness has more specific characteristics and needs more specific explanations. The second explanation is that to enter into another's mind, one must imagine physically entering his body, and a higher-status person finds entering a lower-status person's body repulsive. A third explanation is that clueless people rely upon and invest more in social status because it provides literal meaning in complicated situations; people not naturally talented in strategic thinking gravitate toward status-mediated interactions, such as those within hierarchical organizations, because they need the explicit structure that status provides. A fourth explanation is that cluelessness can improve one's bargaining position; by being visibly clueless, you can commit yourself to not respond to the actions of others. A fifth explanation is that even though strategic thinking is not the same as empathy, perhaps entering another's mind inevitably leads toward empathy, which might threaten the basis of unequal social systems like slavery. Finally, to illustrate the relevance of cluelessness in the real world, I apply these explanations to the U.S. attack on Fallujah in 2004.

CLUELESSNESS IS EASIER

Even though strategic thinking is a routine skill, you can get lazy; the simplest explanation for cluelessness is that strategic thinking takes effort. Sometimes people need to be prodded into thinking strategically; in one experiment, "some subjects do not respond to any consistent set of beliefs when they play a game for the first time and only when they are asked to state beliefs do they form a theory of mind about the opponent" (Costa-Gomes and Weizsäcker 2008, p. 752). Perhaps cluelessness, not strategic thinking, should be considered the default. A college student typically believes that her own actions depend on her own intentions and desires while her roommate's actions depend more on personality and history (Pronin and Kugler 2010). Adults have better theory of mind skills than children, but adults and children are equally egocentric when first hearing another person's request; the difference is that adults are better at subsequently correcting themselves to take into account the requester's knowledge. In other words, "egocentrism isn't outgrown so much as it is overcome each time a person attempts to adopt another's perspective" (Epley, Morewedge, and Keysar 2004, p. 765).

Given the effort involved, perhaps people think strategically only if they have to. In another experiment, asking a person simply to recall an incident in which she had power over others makes her worse at adopting others' points of view and detecting others' emotions. The interpretation is that people in power are "less dependent on others. Thus, to accomplish their goals, the powerful do not need to rely on an accurate, comprehensive understanding of others." Powerful people have "increased demands on [their] attention, so that it is difficult for power holders to take the perspective of everyone under their charge" (Galinsky, Magee, Inesi, and Gruenfeld 2006, p. 1068). College graduates and people who consider themselves of higher social class are generally not as good at detecting others' emotions as high school graduates and people who consider themselves lower class. The interpretation is that "lower-class individuals' life outcomes . . . are more dependent on forces in the external social context. . . . Because of this increased dependence, lower-class individuals tend to focus their attention disproportionately on the context and, in particular, on other people" (Kraus, Côte, and Keltner 2010, p. 1717). In other words, only people buffeted by the actions of others need to spend the mental effort to think strategically.

However, often a clueless person clearly cares a great deal about the outcome. Lady Catherine urgently wants to prevent a marriage between Elizabeth and her nephew Mr. Darcy, and has plenty of time and energy to think about how; her surprise visit to Elizabeth is an active "show of force." General Tilney's failure to anticipate that kicking out Catherine Morland would generate Henry Tilney's sympathy does not result from too many demands on his attention; he goes out of his way to ritually expel her. There are many kinds of mental laziness, such as not being able to stay on a diet (if I am absorbed with work, I don't have the mental energy to choose healthy foods), but cluelessness seems to have particular characteristics, for example, its connection with numeracy and literality. Perhaps a wealthy person can afford to remain clueless about others, but this does not explain why a wealthy person heavily invested in his own status (for example Sir Walter Elliot) is more clueless than a wealthy person who cares little about his own status (for example Mr. Knightley). So although the mental laziness explanation is reasonable, we should consider other more specific explanations.

DIFFICULTY EMBODYING LOW-STATUS OTHERS

When Robert McNamara was appointed U.S. Secretary of Defense, he brought with him quantitative techniques developed at the RAND Corporation, including systems analysis and cost-benefit analysis (Amadae 2003). Game theory was also a big part of RAND at the time, in the late 1940s and early 1950s. RAND economists developed cost-benefit analysis while RAND mathematicians separately developed game theory, according to Leonard (2010, p. 297, note 6).

In any case, not much game theory seems to have rubbed off on McNamara, as one of the life lessons he offers at the end of his career is a game theory truism: when the United States thinks about its enemies, "we must try to put ourselves inside their skins and look at us through their eyes, just to understand the thoughts that lie behind their desires and their actions.... In the Cuban Missile Crisis, at the end, I think we did put ourselves in the skin of the Soviets. In the case of Vietnam, we didn't know them well enough to empathize, and there was total misunderstanding as a result" (Morris 2003). For example, when the U.S. ship *Maddox* was attacked by the Vietnamese on August 2, 1964, the United States assumed that the act must have been ordered by the central command with the intent to escalate, but the attack was in fact ordered by a field commander under standing orders. As McNamara tells Nguyen Dinh Uoc in 1997, "There was a far greater de-centralization of authority and command with respect to the North Vietnamese military than we understood at the time.... [W]e may have drawn unwarranted conclusions, based on our misunderstanding of your command and control arrangements" (Blight and Lang 2005, p. 108; Brams [2011] argues that Jimmy Carter similarly misperceived Ayatollah Khomeini's preferences in the Iran hostage crisis). In his later years, McNamara participated in several conferences with former Cuban, Soviet, and Vietnamese leaders to reach mutual understanding, at least in retrospect, and his efforts are chronicled and celebrated by Blight and Lang (2005).

Why could the U.S. leadership think strategically about the Soviets (thus preventing a catastrophic nuclear exchange) but not the Vietnamese? Social distance is one possible explanation: U.S. leadership at the time included Llewellyn "Tommy" Thompson, who actually lived with Khruschev for a while and knew him socially, while U.S. leaders did not have similar social contact with Vietnamese leadership. The United

States misinterpreted the attack on the *Maddox* as an escalation because of unfamiliarity with North Vietnamese military protocol. Austen would say that without a brother as informant, a woman does not know how little men care about new gowns.

But McNamara's relative inability to put himself in the place of Vietnamese (as opposed to Soviet or Cuban) leaders persists decades later, after plenty of social contact. In a 1989 conference on the Cuban Missile Crisis, McNamara declares, "I want to state quite frankly that with hindsight, if I had been a Cuban leader, I think I might have expected a U.S. invasion" because of the Bay of Pigs invasion and other U.S.-sponsored covert activity, but "we had *absolutely no intention* of invading Cuba...therefore the Soviet action to install missiles...was, I think, based on a misconception—a clearly understandable one, and one that we, in part, were responsible for." Jorge Risquet, a member of Cuba's Politburo at the time, replies, "I am amazed at Mr. McNamara's frankness in acknowledging that if he had found himself in the Cubans' shoes, the Cubans had every right to think that there could be a direct invasion by the Americans" (Blight and Lang 2005, pp. 41–42).

McNamara's overtures toward Vietnamese military leaders are not met with similar regard. In 1995, McNamara met with Vo Nguyen Giap, the senior military strategist during the war. McNamara says, "General, I want us to examine our mindsets, and to look at specific instances where we—Hanoi and Washington—may each have been mistaken, have misunderstood each other," and Giap responds, "I don't believe we misunderstood you. You were the enemy; you wished to defeat us— to destroy us. So we were forced to fight you" (Blight and Lang 2005, p. 105). In 1997, McNamara asks, "My belief is that there could have been negotiations between the end of '65 and '68 which would have led to a settlement that was roughly the same as the one that eventually occurred, but without that terrible loss of life. Why didn't it occur? Were you not influenced by the loss of lives? Why didn't it move you toward negotiations?" Tran Quang Co replies, "[I]f Mr. McNamara thinks that the North Vietnamese leadership was not concerned about the suffering of the Vietnamese people, with deaths and privation, then he has a huge misconception of Vietnam" (Blight and Lang 2005, pp. 52–53).

McNamara's inability to put himself in the minds of Vietnamese leaders persists in his social ineptitude. It should be obvious that the question he poses to Tran Quang Co can be understood as an accusation that the Vietnamese leadership did not care about its own people. McNamara's question is a bit like asking, "Why didn't you give me your wallet earlier? I wouldn't have had to beat you up." McNamara's statement to Jorge Risquet was well received because he admits his own mistake. In contrast, McNamara seems to think that if Vietnamese leaders do not admit their own mistakes, he should be able to get them to do so.

Similarly, Blight and Lang (2005, p. 104) describe Vo Nguyen Giap as "[s]upremely confident, dismissive of any suggestion that any of Hanoi's decisions may have been based on false assumptions about the Americans." They call Giap the "anti-McNamara" and say that "McNamara is unable to convince Giap to keep an open mind." Blight and Lang think that the problem is Giap's, and more generally Vietnam's. They argue that many Vietnamese are critical of the war but cannot voice their criticisms because of suppression; for example, Bao Ninh was placed under house arrest for writing a novel in 1993 that "rejects Giap's smug triumphalism." Blight and Lang deem Giap closed-minded and smug for not adopting McNamara's more self-critical stance. But this is an odd conclusion given their overall advocacy of empathic understanding (if you misunderstand someone, it's your problem, not his). Blight and Lang argue that one should never presume to know another's mind, but when facing someone who doesn't play along, they have no problem thinking that something is wrong with him.

Blight and Lang understand Giap's uncooperative response to McNamara's question not by placing themselves in Giap's shoes, but by saying that he is a certain kind of person: smug, dismissive, and closed-minded. Calling Giap the "anti-McNamara" is also an example of excessive self-reference.

For Blight and Lang, as well as McNamara, status distinctions persist. Lyndon Johnson had offered to stop bombing Vietnam if the Hanoi government stopped supporting insurgents against the U.S.-backed regime in Saigon, but Ho Chi Minh rejected this offer, insisting that the United States unilaterally cease bombing and withdraw its troops before negotiations took place. Blight and Lang (2005, p. 45) interpret this refusal in terms of relative status: "[O]ne of the poorest, most backward countries in the world demands the surrender of the world's greatest superpower. No wonder Johnson was mystified." Instead of trying to understand his adversary's point of view, a clueless superior wonders why the inferior isn't doing what he is supposed to do, just as Lady Catherine does not understand why simply declaring her superior status to Elizabeth is not enough to make her promise not to marry Mr. Darcy, just as Fox does not understand how Flossie could not know he is a fox, and just as the United States is puzzled that its "shock and awe" demonstration of air superiority does not make Iraqi forces simply give up like they are supposed to (Sepp 2007).

What explains McNamara's persistent inability to understand the Vietnamese leadership? McNamara states that to understand someone you have to put yourself in their "skin." In addition to rich versus poor, or advanced versus backward, perhaps another status difference is a racial or "caste" difference: Soviet skin is similar enough for the United States to put itself in while Vietnamese skin is not.

Entering another person's mind often involves imagining yourself physically embodying that person. As mentioned in chapter 2, a young automobile designer at Nissan puts on an "aging suit" to better understand the perspectives of older people. Adam Smith (1759 [2009], pp. 13–14) argues that when we sympathize with another human being, "[b]y the imagination we place ourselves in his situation, we conceive ourselves enduring all the same torments, we enter as it were into his body." Edgar Allan Poe (1845 [1998], p. 258) writes, "When I wish to find out how wise, or how stupid, or how good, or how wicked is any one, or what are his thoughts at the moment, I fashion the expression of my face, as accurately as possible, in accordance with the expression of his, and then wait to see what thoughts or sentiments arise in my mind or heart, as if to match or correspond with the expression" (although Swirski [1996, p. 79] doubts Poe's success with this technique, given his gambling debts). Reading *Persuasion* in 1818, Maria Edgeworth writes, "don't you see Captain Wentworth, or rather don't you in her place feel him taking the boistrous child off her back" (quoted in Southam 1968, p. 17).

Thus one explanation for cluelessness is that physically embodying another person is distasteful, repulsive, even unthinkable. The slave-owner in the "Malitis" tale, who regards diseased meat as proper for slaves, does not consider that slaves might be tricking him; he cannot put himself in the shoes of a slave, cannot look through her eyes, cannot put himself in her skin without discomfort or even revulsion. Under this explanation, Lady Catherine is clueless about Elizabeth not just because entering into her mind would be an admission of status similarity, but because she finds it distasteful to think of herself physically entering into Elizabeth's lower-caste body. When Portuguese explorers went to Africa in the fifteenth century, instead of learning the various African languages themselves and thereby entering into the minds of Africans, they enslaved Africans and brought them back to Portugal to learn Portuguese so they could be used as interpreters in subsequent journeys (Hein 1993).

More recently, Karl Rove, a close advisor to George W. Bush, stated, "Conservatives saw the savagery of 9/11 and the attacks and prepared for war; liberals saw the savagery of the 9/11 attacks and wanted to prepare indictments and offer therapy and understanding for our attackers.... Conservatives saw what happened to us on 9/11 and said, 'We will defeat our enemies.'... Liberals saw what happened to us and said, 'We must understand our enemies' " (Hernandez 2005). Of course, a strategic thinker would say that to defeat an enemy, one should first understand him; for example Sun-tzu (2009, p. 147) writes, "He who knows self/But not the enemy/Will suffer one defeat/For every victory." Using the term "savagery" is a way of saying that the 9/11 attackers are of lower status than civilized human beings, and that the United States debases itself by entering into their minds. In Rove's view, trying to understand the 9/11 attackers should be prohibited, even if understanding them helps defeat them.

For another example, we can understand African American labor resistance as including not just strikes but also more informal actions, such as pilferage, leaving work early, and damaging workplace equipment: "There is evidence of household workers scorching or spitting in food, damaging kitchen utensils, and breaking household appliances, but employers and white contemporaries generally dismissed these acts as proof of black moral and intellectual inferiority...the 'servant problem'" (Kelley 1993, pp. 91, 93). White employers understood the "servant problem" not as a strategy of resistance or a response to insufficient incentives but as a racial characteristic. White employers did not think how they would themselves act in the same situation because being in the same situation was inconceivable. W.E.B. DuBois notes, "All observers spoke of the fact that the slaves were slow and churlish; that they wasted material and malingered at their work. Of course they did. This was not racial but economic. It was the answer of any group of laborers forced down to the last ditch. They might be made to work continuously but no power could make them work well" (DuBois 1935 [1998], p. 40, quoted in Kelley 1993). John Stuart Mill similarly writes, "[T]he most vulgar [of explanations] is that of attributing the diversities of conduct and character to inherent natural differences. What race would not be indolent and insouciant when things are so arranged, that they derive no advantage from forethought or exertion?" (Mill 1848, p. 375, quoted in Levy 2001, p. 95).

INVESTING IN SOCIAL STATUS

In the graphic novel *With the Light* (Tobe 2008, p. 363), Sachiko Azuma describes how her son Hikaru, a child on the autistic spectrum, could not play tag with other children. Hikaru had difficulty understanding how the role of "it" changes: when the "it" person tags someone, that person becomes the new "it." His teacher Mr. Aoki came up with the idea that the person who is "it" should wear a mask, and when that person tags someone, the new "it" puts on the mask. With this small change, Hikaru can join in.

As is true for all social roles, the person who is "it" is defined by a social process. A "police officer" is a person who has gone through specific training, has passed certain specific tests, and is employed by a certain organization. This is a social process, and no person is inherently a police officer, just as no person is inherently "it." In a very small village, I might know the history and experiences of every single individual, and know that a person is a police officer in the same way that I know who "it" is when playing tag. But in larger societies, almost everyone would have the same difficulty as Hikaru. Thus a police officer wears a uniform, and even though we all know that a person becomes a police officer through a lengthy social process, in a given situation it is convenient to think that a person is a police officer because she wears a uniform, as mentioned in chapter 12. The uniform makes the police officer's identity literal.

Social roles and status create literal meaning in social life, as discussed in chapter 12. They provide definition and structure in situations that otherwise might go off in all sorts of directions: with social roles, you don't have to worry about whether the police officer who subdues your robber will invite you to join his church, whether your doctor will ask you for help finding a boyfriend, or whether the person who is "it" will suddenly start to jump rope. Social roles and status create stable, well-defined expectations about what other people will do.

Thus people like Hikaru or Mr. Collins who are not good at strategic thinking, not good at predicting on the fly what other people will do, rely on social status more. They prefer social environments that are structured and defined in terms of status, and prefer to invest their time, energies, and identities in such environments. Under this explanation, a person like Mr. Collins is not clueless because he is status-conscious; rather, because he is clueless, he must rely on social status more and thus invests himself in social environments that are status-mediated and have explicit "rules," such as his formality-filled relationship with his patron, Lady Catherine, and his profession as clergyman, in which he "perform[s] those rites and ceremonies which are instituted by the Church of England" (PP, p. 70).

In other words, stereotypically "male" organizations like the military might be more hierarchical and status-oriented, with each person given an explicit rank, not because men love hierarchy but because their relative cluelessness requires that every social interaction have explicitly defined roles and rules. Mr. Collins might easily call upon his strategic reasoning in well-defined situations like backgammon, in which rules, objectives, and players are explicit, but have more difficulty deploying it in the more open-ended situations typical of social life, and therefore seek the additional definition that social status provides.

African American women in 1960s civil rights organizations such as the SCLC (Southern Christian Leadership Conference) and SNCC (Student Nonviolent Coordinating Committee) were excluded from formal leadership positions (offices such as president or vice-president) because of sexism. The only titled positions available to women were clerical. Thus women who desired more autonomy avoided formal positions and instead became what Robnett (1996) calls "bridge leaders," operating without formal titles and "bridging" the gap between formal leadership and rank-and-file participants.

For example, when the Freedom Riders were almost beaten to death in Alabama in May 1961, Diane Nash called SCLC leader Rev. Fred Shuttlesworth to say that "[t]he students have decided that we can't let violence overcome. We are going to come into Birmingham to continue the Freedom Ride." Shuttlesworth responded, "Young lady, do you know that the Freedom Riders were almost killed here?" Nash replied, "Yes. That's exactly why the rides must not be stopped. If they stop us with violence, the movement is dead. We're coming. We just want to know if you can meet us" (Branch 1988, p. 430, quoted by Robnett 1996, pp. 1685–86). This example illustrates both Nash's strategic acumen and also how as a bridge leader she represented the wishes of the student participants to the formal leadership. Nash also "cement[ed] the policy of staying in jail rather than accepting bail," a crucially effective tactic (Robnett 1996, p. 1686). Bridge leaders in the civil rights movement operated autonomously, creating their own opportunities, caring more about effectiveness than recognition.

Women became bridge leaders because they were excluded from formal leadership. But even in the absence of exclusion, one might expect that people who are good at tactics, recognizing strategic opportunities, and interacting with a wide range of people in a variety of unstructured settings do not need the structure, and constraint, of status-mediated interactions. They might prefer the freedom of operating as an untitled bridge leader over the stable expectations of operating as a formal leader in a status hierarchy. Some people like to jump rope while playing tag.

IMPROVING YOUR BARGAINING POSITION

During the U.S. military occupation of Iraq, many misunderstandings occurred at military checkpoints, resulting in civilian deaths. Ciezadlo (2005) describes the experience:

> As an American journalist here, I have been through many checkpoints and have come close to being shot at several times myself. I look vaguely Middle Eastern, which perhaps makes my checkpoint experience a little closer to that of the typical Iraqi. Here's what it's like.

You're driving along and you see a couple of soldiers standing by the side of the road—but that's a pretty ubiquitous sight in Baghdad, so you don't think anything of it. Next thing you know, soldiers are screaming at you, pointing their rifles and swiveling tank guns in your direction, and you didn't even know it was a checkpoint.

In situations like this, I've often had Iraqi drivers who step on the gas. It's a natural reaction: Angry soldiers are screaming at you in a language you don't understand, and you think they're saying "get out of here," and you're terrified to boot, so you try to drive your way out.

If it's confusing for me—and I'm an American—what is it like for Iraqis who don't speak English?

Another problem is that the U.S. troops tend to have two-stage checkpoints. First there's a knot of Iraqi security forces standing by a sign that says, in Arabic and English, "Stop or you will be shot." Most of the time, the Iraqis will casually wave you through.

Your driver, who slowed down for the checkpoint, will accelerate to resume his normal speed. What he doesn't realize is that there's another American checkpoint several hundred yards past the Iraqi checkpoint, and he's speeding toward it. Sometimes, he may even think that being waved through the first checkpoint means he's exempt from the second one. . . .

A couple of times soldiers have told me at checkpoints that they had just shot somebody. They're not supposed to talk about it, but they do. I think that the soldiers really needed to talk about it. They were traumatized by the experience.

This is not what they wanted—*really* not what they wanted—and the whole checkpoint experience is confusing and terrifying for them as well as for the Iraqis. . . .

The essential problem with checkpoints is that the Americans don't know if the Iraqis are "friendlies" or not, and the Iraqis don't know what the Americans want them to do.

I always wished that the American commanders who set up these checkpoints could drive through themselves, in a civilian car, so they could see what the experience was like for civilians. But it wouldn't be the same. They already know what an American checkpoint is, and how to act at one—which many Iraqis don't.

Given that Iraqi civilians are killed and U.S. soldiers are traumatized, one might expect that U.S. commanders would try hard to communicate to Iraqis approaching a checkpoint what they are being asked to do. As Wright (2004, p. 215) explains, "The question is: Do the Iraqis understand what's going on? . . . [W]arning shots are simply a series of loud bangs and flashes. It's not like this is the international code for 'Stop your vehicle and turn around.'" Instead, the U.S. military developed a laser that temporarily blinds or "dazzles" drivers, in order to incapacitate, not communicate (MSNBC.com 2006). The possible misunderstandings are obvious to low-ranking soliders at the checkpoints (see also McFate 2005), but commanders do not try to resolve them.

Ciezadlo argues that the problem is that commanders do not understand a checkpoint from the point of view of an Iraqi civilian, and offers an explanation for this cluelessness: since commanders already know what a U.S. checkpoint is, they cannot take the point of view of someone who doesn't. But U.S. soldiers operating the checkpoint also already know what a U.S. checkpoint is, and they realize the problem. Perhaps the cluelessness of U.S. commanders is due to revulsion or status preservation: a U.S. soldier perhaps can place himself in the mind of an approaching driver, but a U.S. commander might have difficulty in crossing the status boundary and thinking of himself as an Iraqi civilian driver or even a low-ranking U.S. soldier.

Another explanation, however, is that a U.S. commander can indeed take an Iraqi driver's point of view but does not want to because of perceived risks or costs. To make sure that Iraqi civilians truly understand what to do, one would have to talk with them in close proximity, which might place U.S. soldiers at risk. Once in conversation, a driver might try to bargain, gain familiarity, or plead based on special circumstances, and it would be difficult, especially for a soldier with limited language ability, to figure out which requests are valid. Once you open up the situation to conversation, you open it up to bargaining. As an occupying power, the United States might think that its own interests are best served by keeping an unswerving policy, regardless of the cost in lives. Listening to Iraqi civilians is an admission that Iraqis have legitimate concerns, and the United States might fear that once it starts accommodating them, the inevitably revised policy would place its soldiers at greater risk.

In other words, another explanation for cluelessness is that placing yourself in the mind of another puts you in a worse bargaining position. Cluelessness can serve as a "commitment device," a means of committing yourself not to respond to another, and this can be an advantage in bargaining. For example, in the game of "chicken," two people drive their cars toward each other, and each person can either swerve off the road or go straight. If both go straight, then a terrible accident results; if both swerve, then both emerge unscathed but neither "wins." If one person goes straight and the other swerves, then the one who swerves loses and the one who goes straight wins. This game can be applied to many situations, such as arms races, in which each side would like to build a weapons system, but disaster results if both sides build it.

When playing chicken, say you put a highly visible lock on your steering wheel, making swerving impossible. Seeing the lock, your opponent swerves because she knows you cannot. Thus you gain by deliberately eliminating your options and committing yourself, as illustrated in the movie *Footloose* (Ross 1984), in which Ren McCormack cannot stop his tractor or jump off when his shoelace gets caught in the pedal.

Being visibly clueless is a way to commit oneself. I go straight and presume you will swerve; for me to consider swerving, I have to think about you making a decision to go straight instead. But I make clear to you that I don't think about you making a decision. You will swerve, just like you are supposed to, and my presumption carries the day.

For a related example, say that I approach an intersection in my car (in Los Angeles, where people drive on the right) and am waiting for the oncoming traffic to stop so I can turn left. The traffic light turns from green to yellow, and you approach in the oncoming traffic. If I make my left turn, you will have to stop in order to avoid an accident. If I let you enter the intersection, then I cannot make my left turn and will have to wait another two minutes for the light to turn green again. Thus I should make the turn avoiding eye contact with you, making it clear that I am presuming that you will stop. If I look at you, then you know that I might change my action given what you do.

Under this explanation, a person is clueless because it is a bargaining disadvantage to enter the mind of another, not because of any difficulty violating a status boundary. When I make a left turn while avoiding eye contact with you, I do not have any revulsion about entering into your mind; rather, I do not want to enter your mind because once I start thinking about what you will do, I might hesitate in making my turn.

There is a connection between this bargaining advantage explanation and status. If I can take an action that affects you, you have to think about what I will do and enter into my mind, but if nothing you do affects me, then I don't have to think about you at all. I am clueless about you because I do not have to spend any time thinking about you; it does not even occur to me that you can do anything to hurt or help me, and my cluelessness is a statement that my status is in this sense superior. For example, when police in Birmingham, Alabama, used water hoses and dogs on children demonstrating in the civil rights movement in May 1963, and pictures appeared in newspapers throughout the world, the U.S. federal government was concerned about the image of the U.S. abroad. However, George Wallace, governor of Alabama, made the following statement: "It seems to me that other parts of the world ought to be concerned about what we are thinking of them instead of what they think of us.... After all, we're feeding most of them.... [U]ntil they reject [the money].... that southerners pay for foreign aid to these countries, I will never be concerned about their attitude.... [T]he average man in Africa or Asia doesn't even know where he is, much less where Alabama is" (quoted in Williams 1987, pp. 191–93). Here Wallace says that southerners do not need to care about the minds of people in other parts of the world; we can be clueless about them, but since we give them money, they must be careful about what we think about them. For Wallace, this status difference is naturally linked to race or nationality.

A senior advisor to George W. Bush told Suskind (2004) that reporters like him are "in what we call the reality-based community," people who "believe that solutions emerge from your judicious study of discernible reality." But "[t]hat's not the way the world really works anymore.... We're an empire now, and when we act, we create our own reality. And while you're studying that reality—judiciously, as you will—we'll act again, creating other new realities, which you can study too, and that's how things will sort out. We're history's actors... and you, all of you, will be left to just study what we do." Suskind (2004, p. 51) writes that Bush "clearly feels that unflinching confidence has an almost mystical power." But one need not resort to mysticism; if the Bush administration does not care about what others think, cluelessly creating its own reality, then it has an advantage in some bargaining situations.

George W. Bush told Draper (2007) that "[t]he job of the president...is to think strategically so that you can accomplish big objectives....Iran's a destabilizing force. And instability in that part of the world has deeply adverse consequences, like energy falling in the hands of extremist people that would use it to blackmail the West....That's what I mean by strategic thought. I don't know how you learn that.... [H]ow do you decide, how do you learn to decide things? When you make up your mind, and you stick by it.... You either know how to do it or you don't." Drezner (2007) observes that "what [Bush] has not done is contemplate...[h]ow the Iranian leadership might respond to U.S. policies.... Part of strategic thought is contemplating how others might react to what you do. There's none of that in George W. Bush's strategic thought." Indeed, for Bush strategic thought means making up your mind and sticking to it, and being able to do it is like having a certain status: you either have it or you don't, it's not something you can learn. But again, Bush's version of strategic thought, which does not consider how others react, can be advantageous.

Finally, to take a more juvenile example, I organized the string section for my son's fifth-grade graduation performance and had to assign students to either first violin or second violin. Even though the parts were equally difficult, people often associate first violin with higher status. I assigned Georgia to the second violin part, but she told me that she would rather play first violin, because she was better than Matthew, who was playing first violin. However, she then told me that if she switched with Matthew, then Matthew would get upset, so she would stick with playing second violin. I am confident that this line of strategic reasoning did not occur to Matthew. A typical fifth-grade boy might not even understand this line of reasoning if it were explained to him. Claiming first violin or second violin is an example of chicken: both would rather play first violin but a conflict results if both claim it. A person like Matthew, who does not think about Georgia at all, has an advantage over a person like Georgia, who thinks about what Matthew might do. Amy Heckerling adapted *Emma* into the movie *Clueless* (1995; see also Heckerling 2009), in which the bubbly and scheming Beverly Hills high school student Cher plays the role of Emma. Wald (2000, pp. 229–30) observes: "Being clueless means that Cher is spared the burden of critical self-consciousness that falls to subjects who cannot . . . take for granted a certain freedom of self-expression. . . . It also invests her with an aura of gendered innocence that she can draw upon in negotiations." Matthew's lack of self-consciousness, which helps in his negotiations, is part of being a young boy.

EMPATHY PREVENTION

Another explanation for cluelessness is empathy prevention: if you start thinking about another person's goals and thoughts, you might start to care about them. McNamara's promotion of empathy in international relations draws from White (1984, p. 160, quoted in Blight and Lang 2005, p. 28), who writes, "Empathy is the *great* corrective for all forms of war-promoting misperception. . . . It [means] simply understanding the thoughts and feelings of others. It is distinguished from sympathy. . . . Empathy with opponents is therefore psychologically possible even when a conflict is so intense that sympathy is out of the question." White and McNamara use "empathy" to mean what I call "strategic thinking" and are careful to distinguish it from "sympathy," perhaps because they are close enough that one might lead to the other.

For example, you might have no trouble putting yourself in your opponent's shoes but still would rather not know everything going through his mind. When Indianapolis Colts quarterback Peyton Manning threw an interception to New Orleans Saints defensive back Tracy Porter late in the 2010 Super Bowl, guaranteeing a Saints win, many people speculated that Manning subconsciously wanted to let the Saints win the game (see Moore 2010 for example). Manning's father played ten seasons with the Saints, and Manning himself was born and raised in New Orleans, which in 2010 was still recovering from Hurricane

Katrina and desperate for good news. If you were Manning, any intimate knowledge of the Saints could be very useful in defeating them. But if your knowledge includes knowing just how badly the Saints and New Orleans need the win, you would probably rather not know anything. To take another example, say that you are playing a tennis match and your mother died six months ago. Your opponent's mother died last week and is dedicating this match to her. You would probably rather not know this because your sympathies would make you less competitive, even though this knowledge might be a tactical advantage. Putting yourself in your opponent's shoes is not repulsive; rather, you resist it precisely because you anticipate having too much sympathy for your own good.

Under this explanation, slaveowners do not think strategically about slaves because doing so would inevitably lead to caring about them, which would weaken the social divisions upon which the entire economic system of slavery is based. The preservation of such a system is important enough to outweigh any short-term costs of cluelessness: it is alright for slaves to trick you into giving them free meat occasionally because if you understood them well enough to outwit them, you would no longer believe in slavery. If you place yourself in the minds of Iraqi civilians approaching a checkpoint, you might question whether you should even be in Iraq, a doubt that is more tolerable in an enlisted soldier than an officer.

Singer and Fehr (2005, p. 343) note the tradeoff: "Your capacity to empathize, that is, to simulate the internal state resulting from being cheated in a social exchange will help you to predict the opponent's likely action. Thus, the ability to empathize is useful from a self-interested point of view. However, the very ability to empathize may also undermine purely self-interested choices and may promote other-regarding behavior."

CALLING PEOPLE ANIMALS

On March 31, 2004, four U.S. military contractors working for Blackwater Security Consulting were ambushed and killed by Iraqi insurgents in Fallujah; their bodies were mutilated and hung from a bridge. The *New York Times* reported that "Enraged Mob in Falluja Kills 4 American Contractors" (Gettleman 2004). On April 1, L. Paul Bremer, head of the Coalition Provisional Authority in Iraq, made the following statement during the commencement ceremony for new graduates of the police academy in Baghdad: "Yesterday's events in Fallujah are a dramatic example of the ongoing struggle between human dignity and barbarism. Five brave soldiers were killed by an attack in their area.

Then, two vehicles containing four Americans were attacked and their bodies subjected to barbarous maltreatment.... You, the finest of the honorable majority of Iraq's men and women, have chosen to confront the evil-doers, to carry the banner of civilization. You have put to shame the human jackals who defiled the streets of Fallujah" (Bremer 2004).

On the evening of April 4, U.S. military forces surrounded Fallujah and on April 5 began attacking insurgent positions, also killing hundreds of civilians. On September 13, Lt. General James T. Conway, the top U.S. general in charge of western Iraq, told reporters that he had opposed the Fallujah invasion: "We felt...that we ought to probably let the situation settle before we appeared to be attacking out of revenge" (Chandrasekaran 2004). According to a UPI report, "[S]enior commanders universally said in interviews...[that] they would not have gone into Fallujah at that time under those conditions.... [T]he immediate aftermath of the March 31 killings was not the time to fight, they said. First, that robbed them of the element of surprise.... Second, it 'taught' the insurgents that their provocative acts could draw the United States into an urban battle when they wanted it, rather than the other way around. Third, finding the individuals whose faces were on the videotape of the contractor killings is in essence a police job.... [I]t would be easy under peaceful conditions to have local police find the identities of the killers and arrest them" (Hess 2004). However, perceiving a challenge to the superior status of the United States, George W. Bush stated, "We will not be intimidated.... We will finish the job" (Rubin and McManus 2004).

The attack on Fallujah was a public relations disaster for the United States. "[A]s thousands of Fallujans escaped their city and fled to other parts of Iraq, they brought with them tales of horror and civilian death that no amount of propaganda could combat" (Scahill 2008, p. 205). Indeed, U.S. cluelessness is evident in its obsession with status. Insurgents have a subhuman status; Bremer uses caste language like "defile." The purpose of the Iraqi police graduation ceremony is to establish the new social status of the graduates, and even though they have nothing to do with the Fallujah killings, Bremer feels free to use them and their ceremony to heighten the status difference between them and the insurgents, between "good" and "evil" Iraqis. In contrast, U.S. military leaders knew that the operation had fundamental tactical flaws. U.S. commanders had preferred to understand the issue in terms of capturing individual criminals, not a battle between one kind of people (guardians of civilization) and another kind (enraged mob). Lakhdar Brahimi, United Nations envoy to Iraq, called the U.S. attack "collective punishment" (Rubin and McManus 2004).

Bremer's terminology is quite specifically animal: the killers are jackals, not just barbarians or savages. If it is difficult to put yourself into the mind (and body) of a barbarian, it is harder still to put yourself into the mind (and body) of a jackal. Calling your enemy an animal diminishes your empathy for him and thereby makes it easier to kill him; elsewhere in his statement Bremer uses the word "ghouls," who are already dead. But in addition, calling your enemy an animal is a way of saying that you don't have to think about his motivations, making, for example, negotiations and bargaining, which require you to acknowledge his goals, inconceivable. According to Vermeule (2010, p. 195), "[S]ituational mind blindness...deny[ing] other people the perspective of rational agency by turning them into animals, machines, or anything without a mind" is a more complex kind of dehumanization than simply positing that "members of the hated countergroup do not count as human, and therefore moral rules do not apply to them."

Calling your enemy an animal might improve your bargaining position or deaden your moral qualms, but at the expense of not being able to think about your enemy strategically. On April 9, due to international outrage and also the military's concern that the attack was creating new recruits for the insurgency, the U.S. announced a unilateral cease-fire (Rubin and McManus 2004). After negotiations, the U.S. established the Fallujah Brigade, supplying over eight hundred assault rifles and over twenty-five trucks so Fallujah residents could govern themselves. Creating a new social status, a new kind of good Iraqi, was insufficient, however, and these weapons and vehicles were soon used in insurgent attacks.

When Austen analyzed cluelessness through the blunders of Lady Catherine and General Tilney, she probably did not anticipate that her explanations might also apply to the blunders of the United States two hundred years later in Iraq. Austen might agree that cluelessness is a real and recurrent phenomenon, worthy of further study.

Concluding Remarks

PEOPLE HAVE been trying to understand humans for a long time. Elster (2007, pp. x, 5–6) writes, "If we neglect twenty-five centuries of reflection about mind, action, and interaction in favor of the last one hundred years or the last ten, we do so at our peril and our loss" (see also Elster 1999). However, Elster does acknowledge recent progress: "rational-choice theory is nevertheless a valuable part of the toolbox.... Game theory, in particular, has illuminated the structure of social interaction in ways that go far beyond the insights achieved in earlier centuries." Leonard (2010, p. 1) writes, "Game theory, it may be reasonably be claimed, has proved to be one of the more significant scientific contributions of the twentieth century."

But game theory goes way back also. Parikh (2008) finds game-theoretic insights in the Akbar and Birbal stories from the sixteenth century and also in the Mahabharata (see also Sihag 2007 and Wiese 2012). O'Neill (1982, 2009) and Aumann and Maschler (1985) consider the problem of conflicting claims to an estate; using game theory, they are able to "reverse engineer" solutions advanced in the Babylonian Talmud more than 1500 years ago. Ober (2011) finds that Greek authors, including Herodotus and Thucydides, "regarded individuals and groups as...capable of planning and acting strategically" and insightfully analyzed, thousands of years ago, how societies induce individuals to cooperate. Vanderschraaf (1998) finds that David Hume, in 1740, elucidated fundamental game-theoretic concepts, such as the idea of an equilibrium outcome.

This book focuses on Austen and African American folktales, but folk game theory can be found in many different places. A splendid example is *Oklahoma!*, the first Rodgers and Hammerstein musical and perhaps the most influential in the history of American musical theater. In *Oklahoma!* the best strategists are women (Ado Annie and Aunt Eller) or ethnic outsiders (the Persian peddler Ali Hakim). Ado Annie, the "girl who can't say no," like Flossie Finley and Catherine Morland, demonstrates the strategic advantages of projecting naivety. Ali Hakim, like Brer Rabbit and Emma Woodhouse, illustrates the perils of overstrategicness. Will Parker, like Flossie's Fox and Sir Walter Elliot, illustrates how overattention to social status and literal meaning indicates strategic imbecility. Laurey, like Fanny Price, illustrates how learning to think strategically is part of becoming a grown woman. Finally, Curly and Laurey, like Austen's couples, show how strategic partnership is a foundation for marriage.

In the musical number "It's a Scandal! It's an Outrage!," Ali Hakim calls for a revolution against the existing courtship regime in which "It's gittin' so you cain't have any fun!/Ev'ry daughter has a father with a gun!" (Hammerstein 1942, p. 18). But when the single men he is organizing ask him who will be the first man to be shot, he realizes that it must be him, as their leader. As the men start their revolution by dancing, swinging round and thereby revolving on the stage, the stage directions state that the girls "pick them off the line and walk off with them" (Hammerstein 1942, p. 19). Long before the social science literature on the free-rider problem (Olson 1965), *Oklahoma!* states the central strategic problem in revolution: no one wants to be the first person to be shot, and any collective action is vulnerable to individuals being picked off one by one. When Ado Annie says to him that because he married Gertie, he must have wanted to marry her, Ali Hakim responds exactly as a rational choice theorist would: "Sure I wanted to. I wanted to marry her when I saw the moonlight shining on the barrel of her father's shotgun!" (Hammerstein 1942, p. 46). Ali Hakim sells the essence of rational choice theory, the "elixir of Egypt," to Laurey, explaining that when "Pharoah's daughter . . . had a hard problem to decide . . . she'd take a whiff of this" (Hammerstein 1942, p. 12). Laurey complies, saying that "it's goin' to make up my mind fer me" while the girls around her sing, "Make up your mind, make up your mind Laurey" (Hammerstein 1942, p. 28).

The first actor to play Ali Hakim on Broadway was the Yiddish actor Joseph Buloff, and in a party celebrating *Oklahoma!*'s first anniversary, Hammerstein was billed as "Mister Ali Hakimstein." Most (1998) speculates that Ali Hakim's name comes from the Yiddish and Hebrew word *hacham*, a "clever man." Thus Hammerstein imports the Jewish folk game theory tradition into the mythology of the American westward expansion. Similarly, graphic novelist Steve Sheinkin (2010) chronicles the adventures of Rabbi Harvey in Wild West Colorado, and Newbery Medal-winning Sid Fleischman (1963 [1988], pp. 22–26) places a lie detection scheme from Elijah ben Solomon, and even earlier from Akbar and Birbal, into the California gold rush (Sheinkin 2010, p. 134; Sarin 2005, p. 32).

We can also learn more from Austen. For example, economic theory is not Austen's main concern, but since economic theory is based upon rational choice theory, it is not surprising that she makes advances in this direction. One of the oldest questions in economics is what determines an object's value. There is a temptation to think that an object's value is inherent in its physical properties or how it was produced. For example, the labor theory of value posits that an object's value depends on the amount of labor that went into producing it; thus a chair has higher value than the raw wood it was made from.

Mr. Knightley thinks Harriet Smith of low value and not truly worthy of Mr. Robert Martin because of her personal attributes: "What are Harriet Smith's claims, either of birth, nature or education, to any connection higher than Robert Martin? She is the natural daughter of nobody knows whom, with probably no settled provision at all, and certainly no respectable relations.... She has been taught nothing useful, and is too young and too simple to have acquired any thing herself.... She is pretty, and she is good tempered, and that is all" (E, p. 64). Emma replies in Harriet's defense: "[S]he is, in fact, a beautiful girl, and must be thought so by ninety-nine people out of an hundred; and till it appears that men are much more philosophic on the subject of beauty than they are generally supposed; till they do fall in love with well-informed minds instead of handsome faces, a girl, with such loveliness as Harriet, has a certainty of being admired and sought after, of having the power of choosing from among many, consequently a claim to be nice" (E, p. 67).

Emma explains that Harriet's value is not determined directly by her attributes but by how much others desire her and what she can extract from the marriage market, given her large number of potential suitors. Austen's theory of value is that the value of a good is determined by what people will exchange for it in market transactions. The value of a good cannot be reduced to its attributes or the labor that went into making it, but depends on the entire context of how the good is exchanged, including how many people want it and how badly, and how many people have it to sell. For example, Jevons (1871, pp. 2, 156) writes, "Repeated reflection and inquiry have led me to the somewhat novel opinion, that *value depends entirely on utility*.... Bread has the almost infinite utility of maintaining life, and when it becomes a question of life or death, a small quantity of food exceeds in value all other things. But when we enjoy our ordinary supplies of food, a loaf of bread has little value." This is the standard view of modern economic theory, which followed Austen by only several decades.

The congruence of narrative and social theory continues today, even in the abstract mathematical models used in economics. Several have pointed out the similarity between economic models and stories (Cowen 2005; Morgan 2010). For example, Rubinstein (2012b, p. x) writes, "The word 'model' sounds more scientific than 'fable' or 'fairy tale', but I don't see much difference between them.... Being something between fantasy and reality, a fable is free of extraneous details and annoying diversions.... [Thus] we can clearly discern what cannot always be seen from the real world. On our return to reality, we are in possession of some sound advice or a relevant argument that can be used in the real world. We do exactly the same thing in economic theory" (see also Rubinstein 2012a). Similarly, Leamer (2012, p. 10) writes, "We would do better theory if we judged economic theories by the same standards that we judge novels." According to Gallagher (2006, p. 192), "[L]iterary critics are now more curious and tolerant about economic logic than they were at any time in the twentieth century."

If game theory can be developed using narratives, why mathematize? Are recent technical innovations necessary? A person with well-developed social skills might see mathematical game theory like Elizabeth Bennet sees whist and backgammon, as overspecialized and irrelevant. Recently the formalism of modern social science (economics in particular) has been called "autistic," as if that's a bad thing (Mohn 2010; Devine 2006). Repetitive obsession with mathematics can certainly hinder real-world understanding, and drawing a game tree to show how Fanny gets Betsey to give up Susan's knife might well be technical overkill for such a simple situation. But at least a game tree is a different kind of representation, one that is visual and concrete. Teachers working with students on the autistic spectrum have developed visual representations, such as cartoons and thought bubbles, to make people's thoughts and motivations explicit (Gray 1994; Baker 2001; Cohen and Sloan 2007). Game theory similarly uses trees, tables, and numbers. According to Cowen (2009, p. 170), Adam Smith (1759 [2009]) writes about sympathy from the perspective of an autistic outsider, who must keenly observe and analyze others because he does not "get it" himself. Thus along with its extensibility, a mathematical perspective is useful precisely because it is different, because it makes one state explicitly what some find obvious. This explicitness might also be the only way to approach more complicated situations that are not obvious to anyone.

With math in mind, recall that Elizabeth and Jane Bennet find their sister "Mary, as usual, deep in the study of thorough bass and human nature; and had some new extracts to admire, and some new observations of thread-bare morality to listen to" (PP, p. 67). As mentioned in chapter 12, thorough bass uses numbers over a bass line to indicate chords, as shown in figure 5 (C.P.E. Bach, in Arnold 1931 [1965], p. 662). Do thorough bass and human nature go together? Rogers (2006, p. 483) says that Austen presents thorough bass as a "comically recondite study, juxtaposed with the all-encompassing 'human nature.' " Similarly, Wallace (1983, p. 10) says that "one refers to music, the other to life; the one to technique, the other to spirit...

Mary is unable to discriminate; she approaches both music and life in the same overzealous and unimaginative way."

Figure 5. Thorough bass: a bass line with numerical figures is in the bottom staff, and the top staff shows the realized chords

What does it mean to approach human nature in the same way as thorough bass? Thorough bass, or figured bass, is a kind of technical notation that remains in music theory today. It is mathematical not only because it uses numbers to abbreviate, but also because musical notation itself is a mathematical, technical achievement that took hundreds of years to develop; according to Crosby (1997, p. 144), "[t]he musical staff was Europe's first graph," long before graphs were used to represent physical phenomena. Also, thorough bass is not just a kind of notation, but an entire way of thinking about music: a keyboard player reading thorough bass would not simply read the chords, but realize them improvisationally, choosing notes and embellishments according to the needs of the ensemble. Christensen (2010, p. 40) writes, "If thorough bass remained in the estimation of most musicians a lowly, practical art, there were clearly many others who saw in it something far greater: an art that at its best requires all the skills and imagination of the composer." Thorough bass was not just useful but generative, enabling the creation of new music and new music theory; for example, Christensen (2010, p. 9) writes that "[t]he pedagogical mnemonics by which figured-bass was taught to young musicians became a surprisingly powerful instigation for remarkable developments in the area of tonal music theory." Indeed, "[t]he seventeenth century, and the greater part of the eighteenth, were known to the majority of historians and critics as the period of the thorough-bass, and it is only in recent years that the term 'baroque' came to be accepted" (Stevens 1965, p. vii). Of course, music can develop without any kind of written or visual representation. But the development of Western music would have been inconceivably different without the technical innovation of music notation.

By having Mary understand human nature and thorough bass in the same way, Austen suggests, however briefly, the possibility of understanding human behavior using a technical, mathematical approach, in particular an approach that prizes the ability to elegantly translate between technical representations and real-world performances. In support of this claim, this is not the only time Austen connects human behavior with music and mathematics. As mentioned in chapter 11, when Mr. Darcy claims that he is not good at introducing himself to strangers, Elizabeth, seated at the piano, says that it is just a matter of taking "the trouble of practising," and Mr. Darcy replies, "We neither of us perform to strangers" (PP, p. 197). When Elizabeth decides to take the long muddy walk to visit Jane when she is ill, Mary observes that "exertion should always be in proportion to what is required" (PP, p. 35).

Defenders of thorough bass have taken issue with Austen; for example, "It is puzzling to find Jane Austen (for it is the authoress herself who speaks here) sneering at a plain girl for studying a branch of musical theory. Why was Mary to be ridiculed for doing something so very well worth while?" (Piggott 1979, p. 54). But Austen's only negative word is "thread-bare," which she counterposes with "new"; "thread-bare" does not mean stupid or foolish but rather worn-out, hackneyed, overused. In other words, what Austen ridicules is not thorough bass or its application to human nature, but the production of boring, well-known, moralistic results. Austen is interested in new results. If Mary Bennet has not found any yet, at least she suggests a direction.

References

Abbott, Andrew. 2004. *Methods of Discovery: Heuristics for the Social Sciences.* New York: W. W. Norton.

Abella, Alex. 2008. *Soldiers of Reason: The RAND Corporation and the Rise of the American Empire.* Boston and New York: Houghton Mifflin Harcourt.

Abelson, Robert P. 1996. "The Secret Existence of Expressive Behavior." In *The Rational Choice Controversy: Economic Models of Politics Reconsidered,* Jeffrey Friedman, editor. New Haven: Yale University Press.

Ainslie, George. 1992. *Picoeconomics: The Strategic Interaction of Successive Motivational States within the Person.* Cambridge: Cambridge University Press.

Amadae, S. M. 2003. *Rationalizing Capitalist Democracy: The Cold War Origins of Rational Choice Liberalism.* Chicago: University of Chicago Press.

Ames, Daniel L., Adrianna C. Jenkins, Mahzarin R. Banaji, and Jason P. Mitchell. 2008. "Taking Another Person's Perspective Increases Self-Referential Neural Processing." *Psychological Science* 19: 642–44.

Anderson, Elizabeth. 1997. "Practical Reason and Incommensurable Goods." In *Incommensurability, Incomparability, and Practical Reason,* Ruth Chang, editor. Cambridge: Harvard University Press.

Archer, Margaret S. 2000. "*Homo economicus, Homo sociologicus,* and *Homo sentiens.*" In *Rational Choice Theory: Resisting Colonization,* Margaret S. Archer and Jonathan Q. Tritter, editors. London: Routledge.

Archer, Margaret S., and Jonathan Q. Tritter, editors. 2000. *Rational Choice Theory: Resisting Colonization.* London: Routledge.

Arnold, F. T. 1931 [1965]. *The Art of Accompaniment from a Thorough-Bass as Practiced in the XVIIth and XVIIIth Centuries.* Mineola, N. Y.: Dover.

Aumann, Robert J., and Michael Maschler. 1985. "Game Theoretic Analysis of a Bankruptcy Problem from the Talmud." *Journal of Economic Theory* 36: 195–213.

Austen, Jane. 1811 [2006]. *Sense and Sensibility.* Edward Copeland, editor. Cambridge: Cambridge University Press.

———. 1813 [2006]. *Pride and Prejudice.* Pat Rogers, editor. Cambridge: Cambridge University Press.

———. 1814 [2005]. *Mansfield Park.* John Wiltshire, editor. Cambridge: Cambridge University Press.

———. 1816 [2005]. *Emma.* Richard Cronin and Dorothy McMillan, editors. Cambridge: Cambridge University Press.

———. 1817 [2006]. *Northanger Abbey.* Barbara M. Benedict and Deirdre Le Faye, editors. Cambridge: Cambridge University Press.

———. 1817 [2006]. *Persuasion.* Janet Todd and Antje Blank, editors. Cambridge: Cambridge University Press.

Austin, J. L. 1975. *How to Do Things with Words*. Second edition. J. O. Urmson and Marina Sbisà, editors. Cambridge: Harvard University Press.

Bailey, Mel. 1985. "Interview with Sheriff Mel Bailey." Conducted by Blackside, Inc., on November 2, 1985, for *Eyes on the Prize: America's Civil Rights Years (1954–1965)*. Washington University Libraries, Film and Media Archive, Henry Hampton Collection. http://digital.wustl.edu/e/eop/browse.html.

Baker, Jed. 2001. *The Autism Social Skills Picture Book: Teaching Communication, Play and Emotion*. Arlington, Texas: Future Horizons.

Bancroft, Lundy. 2002. *Why Does He Do That? Inside the Minds of Angry and Controlling Men*. New York: Berkley Books.

Baron, Jonathan. 2008. *Thinking and Deciding*. Fourth edition. Cambridge: Cambridge University Press.

Baron-Cohen, Simon. 1997. *Mindblindness: An Essay on Autism and Theory of Mind*. Cambridge: MIT Press.

Baron-Cohen, Simon. 2002. "The Extreme Male Brain Theory of Autism." *Trends in Cognitive Sciences* 6: 248–54.

Baron-Cohen, Simon, Therese Jolliffe, Catherine Mortimore, and Mary Robertson. 1997. "Another Advanced Test of Theory of Mind: Evidence From Very High Functioning Adults with Autism or Asperger Syndrome." *Journal of Child Psychology and Psychiatry* 38: 813–22.

Baron-Cohen, Simon, Alan M. Leslie, and Uta Frith. 1985. "Does the Autistic Child Have a 'Theory of Mind'?" *Cognition* 21: 37–46.

Beeger, Sander, Bertram F. Malle, Mante S. Nieuwland, and Boaz Keysar. 2010. "Using Theory of Mind to Represent and Take Part in Social Interactions: Comparing Individuals with High-functioning Autism and Typically Developing Controls." *European Journal of Developmental Psychology* 7: 104–22.

Belenky, Mary Field, Blythe McVicker Clinchy, Nancy Rule Goldberger, and Jill Mattuck Tarule. 1986. *Women's Ways of Knowing: The Development of Self, Voice, and Mind*. New York: Basic Books.

Bender, John. 1987. *Imagining the Penitentiary: Fiction and the Architecture of Mind in Eighteenth-Century England*. Chicago: University of Chicago Press.

———. 2012. "Rational Choice in Love: *Les Liaisons dangereuses*." In *Ends of Enlightenment*, by John Bender. Stanford: Stanford University Press.

Benhabib, Jess, and Alberto Bisin. 2005. "Modeling Internal Commitment Mechanisms and Self-Control: A Neuroeconomics Approach to Consumption-Saving Decisions." *Games and Economic Behavior* 52: 460–92.

Bevel, James. 1985. "Interview with James Bevel." Conducted by Blackside, Inc., on November 13, 1985, for *Eyes on the Prize: America's Civil Rights Years (1954–1965)*. Washington University Libraries, Film and Media Archive, Henry Hampton Collection. http://digital.wustl.edu/e/eop/browse.html.

Bhatt, Meghana, and Colin F. Camerer. 2005. "Self-Referential Thinking and Equilibrium as States of Mind in Games: fMRI Evidence." *Games and Economic Behavior* 52: 424–59.

Binmore, Ken. 2007. *Playing for Real: A Text on Game Theory*. Oxford: Oxford University Press.

Blight, James G., and janet M. Lang. 2005. *The Fog of War: Lessons from the Life of Robert S. McNamara*. Lanham, Md.: Rowman & Littlefield.

Bloom, Paul, and Tim P. German. 2000. "Two Reasons to Abandon the False Belief Task as a Test of Theory of Mind." *Cognition* 77: B25–B31.

Botkin, B. A., editor. 1945. *Lay My Burden Down: A Folk History of Slavery.* Chicago: University of Chicago Press.

Bourdieu, Pierre, and Loïc J. D. Wacquant. 1992. *An Invitation to Reflexive Sociology.* Chicago: University of Chicago Press.

Brams, Steven J. 1994. "Game Theory and Literature." *Games and Economic Behavior* 6: 32–54.

Brams, Steven J. 2002. *Biblical Games: Game Theory and the Hebrew Bible.* Cambridge: MIT Press.

———. 2011. *Game Theory and the Humanities: Bridging Two Worlds.* Cambridge: MIT Press.

Branch, Taylor. 1988. *Parting the Waters: America in the King Years, 1954–1963.* New York: Simon & Schuster.

Bray, Joe. 2007. "The 'Dual Voice' of Free Indirect Discourse: A Reading Experiment." *Language and Literature* 16: 37–52.

Bremer, L. Paul. 2004. Police Academy Commencement Speech. Baghdad, April 1. http://govinfo.library.unt.edu/cpa-iraq/transcripts/20040401_bremer_police.html.

Brownstein, Rachel M. 1997. "*Northanger Abbey, Sense and Sensibility, Pride and Prejudice.*" In *The Cambridge Companion to Jane Austen,* Edward Copeland and Juliet McMaster, editors. Cambridge: Cambridge University Press.

Butte, George. 2004. *I Know That You Know That I Know: Narrating Subjects from Moll Flanders to Marnie.* Columbus, Ohio: Ohio State University Press.

Calhoun, Craig. 2001. "A Structural Approach to Social Movement Emotions." In *Passionate Politics: Emotions and Social Movements,* Jeff Goodwin, James M. Jasper, and Francesca Polletta, editors. Chicago: University of Chicago Press.

Camic, Charles. 1986. "The Matter of Habit." *American Journal of Sociology* 91: 1039–87.

Cassidy, Kimberly Wright, Deborah Shaw Fineberg, Kimberly Brown, and Alexis Perkins. 2005. "Theory of Mind May Be Contagious, but You Don't Catch It from Your Twin." *Child Development* 76: 97–106.

Chandrasekaran, Rajiv. 2004. "Key General Criticizes April Attack in Fallujah: Abrupt Withdrawal Called Vacillation." *Washington Post,* September 13.

Chrissochoidis, Ilias, Heike Harmgart, Steffen Huck, and Wieland Müller. 2010. " 'Though this be madness, yet there is method in't.' A Counterfactual Analysis of Richard Wagner's *Tannhäuser.*" Working paper, University College London.

Chrissochoidis, Ilias, and Steffen Huck. 2011. "Elsa's Reason: On Beliefs and Motives in Richard Wagner's *Lohengrin.*" *Cambridge Opera Journal* 22: 65–91.

Christensen, Thomas. 2010. "Thoroughbass as Music Theory." In *Partimento and Continuo Playing in Theory and Practice,* Dirk Moelants, editor. Leuven: Leuven University Press.

Chwe, Michael Suk-Young. 1990. "Why Were Workers Whipped? Pain in a Principal-Agent Model." *Economic Journal* 100: 1109–21.

———. 2001. *Rational Ritual: Culture, Coordination, and Common Knowledge.* Princeton: Princeton University Press.

———. 2009. "Rational Choice and the Humanities: Excerpts and Folk-tales." *Occasion: Interdisciplinary Studies in the Humanities* 1 (October 15). http://arcade.stanford.edu/journals/occasion/.

Ciezadlo, Annia. 2005. "What Iraq's Checkpoints Are Like." *Christian Science Monitor,* March 7.

Cloud, John. 2009. "Why Exercise Won't Make You Thin." *Time,* August 9.

Cohen, Marlene J., and Donna L. Sloan. 2007. *Visual Supports for People with Autism: A Guide for Parents and Professionals.* Bethesda, Md.: Woodbine House.

Conan Doyle, Arthur. 1893 [2005]. "The Final Problem." In *The New Annotated Sherlock Holmes,* volume 1. Leslie S. Klinger, editor. New York: W. W. Norton.

Cosmides, Leda, and John Tooby. 1994. "Beyond Intuition and Instinct Blindness: Toward an Evolutionarily Rigorous Cognitive Science." *Cognition* 50: 41–77.

Costa-Gomes, Miguel A., and Georg Weizsäcker. 2008. "Stated Beliefs and Play in Normal-Form Games." *Review of Economic Studies* 75: 729–62.

Cowen, Tyler. 2005. "Is a Novel a Model?" Working paper, Department of Economics, George Mason University.

———. 2009. *Create Your Own Economy: The Path to Prosperity in a Disordered World.* New York: Dutton.

Cramer, Christopher. 2002. "*Homo Economicus* Goes to War: Methodological Individualism, Rational Choice and the Political Economy of War." *World Development* 30: 1845–64.

Crawford, Vincent P., Miguel A. Costa-Gomes, and Nagore Iriberri. 2010. "Strategic Thinking." Working paper, Department of Economics, University of Oxford.

Crick, Nicki R., and Jennifer K. Grotpeter. 1995. "Relational Aggression, Gender, and Social-Psychological Adjustment." *Child Development* 66: 710–22.

Crosby, Alfred W. 1997. *The Measure of Reality: Quantification and Western Society, 1250–1600.* Cambridge: Cambridge University Press.

Cziko, Gary A. 1989. "Unpredictability and Indeterminism in Human Behavior: Arguments and Implications for Educational Research." *Educational Researcher* 18: 17–25.

Daston, Lorraine. 2004. "Whither *Critical Inquiry?*" *Critical Inquiry* 30: 361–64.

Deloche, Régis, and Fabienne Oguer. 2006. "Game Theory and Poe's Detective Stories and Life." *Eastern Economic Journal* 32: 97–110.

De Ley, Herbert. 1988. "The Name of the Game: Applying Game Theory in Literature." *SubStance* 17: 33–46.

Devine, James G. 2006. "Psychological Autism, Institutional Autism, and Economics." In *Real World Economics: A Post-Autistic Economics Reader.* Edward Fullbrook, editor. London: Anthem Press.

Dijk, Corine, Peter J. de Jong, and Madelon L. Peters. 2009. "The Remedial Value of Blushing in the Context of Transgressions and Mishaps." *Emotion* 9: 287–91.

Dimand, Mary Ann, and Robert W. Dimand. 1996. *A History of Game Theory, Volume 1: From the Beginnings to 1945.* London and New York: Routledge.

Dixit, Avinash K. 2005. "Restoring Fun to Game Theory." *Journal of Economic Education* 36: 205–19.

Dixit, Avinash K., and Barry Nalebuff. 2008. *The Art of Strategy: A Game Theorist's Guide To Success in Business and Life.* New York: W. W. Norton.

Doody, Margaret Anne. 1988. *Frances Burney: The Life in the Works.* New Brunswick, New Jersey: Rutgers University Press.

Draper, Robert. 2007. *Dead Certain: The Presidency of George W. Bush.* New York: Free Press.

Dreiser, Theodore. 1900 [1981]. *Sister Carrie.* New York: Penguin Books.

Drezner, Daniel. 2007. "The Strategic Thought of George W. Bush." http://www.danieldrezner.com/archives/003479.html.

DuBois, W.E.B. 1935 [1998]. *Black Reconstruction in America.* New York: Free Press.

Duckworth, Alastair M. 1975. " 'Spillikins, paper ships, riddles, conundrums, and cards': Games in Jane Austen's Life and Fiction." In *Jane Austen: Bicentenary Essays,* John Halperin, editor. Cambridge: Cambridge University Press.

Duesenberry, James S. 1960. "Comment." In *Demographic and Economic Change in Developed Countries,* National Bureau of Economic Research. Princeton: Princeton University Press.

Edwards, Jr., Thomas R. 1965. "The Difficult Beauty of Mansfield Park." *Nineteenth-Century Fiction* 20: 51–67.

Eliaz, Kfir, and Ariel Rubinstein. 2011. "Edgar Allan Poe's Riddle: Framing Effects in Repeated Matching Pennies Games." *Games and Economic Behavior* 71: 88–99.

Elster, Jon. 1983. *Sour Grapes: Studies in the Subversion of Rationality.* Cambridge: Cambridge University Press.

———. 1999. *Alchemies of the Mind: Rationality and the Emotions.* Cambridge: Cambridge University Press.

———. 2007. *Explaining Social Behavior: More Nuts and Bolts for the Social Sciences.* Cambridge: Cambridge University Press.

England, Paula, and Barbara Stanek Kilbourne. 1990. "Feminist Critiques of the Separative Model of Self: Implications for Rational Choice Theory." *Rationality and Society* 2: 156–71.

Epley, Nicholas, Carey K. Morewedge, and Boaz Keysar. 2004. "Perspective Taking in Children and Adults: Equivalent Egocentrism but Differential Correction." *Journal of Experimental Social Psychology* 40: 760–68.

Evans-Campbell, Teresa, Karen I. Fredriksen-Goldsen, Karina L. Walters, and Antony Stately. 2007. "Caregiving Experiences among American Indian Two-Spirit Men and Women: Contemporary and Historical Roles." *Journal of Gay and Lesbian Social Services* 18: 75–92.

Ferguson Bottomer, Phyllis. 2007. *So Odd a Mixture: Along the Autistic Spectrum in 'Pride and Prejudice.'* London: Jessica Kingsley.

Finch, Casey, and Peter Bowen. 1990. " 'The Tittle-Tattle of Highbury': Gossip and the Free Indirect Style in *Emma.*" *Representations* 31: 1–18.

Fleischman, Sid. 1963 [1988]. *By the Great Horn Spoon!* New York: Little, Brown.

Fourcade, Marion. 2009. *Economists and Societies: Discipline and Profession in the United States, Britain, and France, 1890s to 1990s.* Princeton: Princeton University Press.

Frank, Robert H. 1988. *Passions within Reason: The Strategic Role of the Emotions.* New York: Norton.

———. 2004. *What Price the Moral High Ground? Ethical Dilemmas in Competitive Environments.* Princeton: Princeton University Press.

Friedman, Daniel, Kai Pommerenke, Rajan Lukose, Garrett Milam, and Bernardo A. Huberman. 2007. "Searching for the Sunk Cost Fallacy." *Experimental Economics* 10: 79–104.

Friedman, Jeffrey, editor. 1996. *The Rational Choice Controversy: Economic Models of Politics Reconsidered.* New Haven: Yale University Press.

Fudenberg, Drew, and David K. Levine. 2006. "A Dual-Self Model of Impulse Control." *American Economic Review* 96: 1449–76.

Galinsky, Adam D., Joe C. Magee, M. Ena Inesi, and Deborah H. Gruenfeld. 2006. "Power and Perspectives Not Taken." *Psychological Science* 17: 1068–74.

Gallagher, Catherine. 2006. *The Body Economic: Life, Death, and Sensation in Political Economy and the Victorian Novel.* Princeton: Princeton University Press.

Gernsbacher, Morton Ann, and Jennifer L. Frymiare. 2005. "Does the Autistic Brain Lack Core Modules?" *Journal of Developmental and Learning Disorders* 9: 3–16.

Gettleman, Jefferey. 2004. "Enraged Mob in Falluja Kills 4 American Contractors." *New York Times,* March 31.

Gilley, Brian Joseph. 2006. *Becoming Two-Spirit: Gay Identity and Social Acceptance in Indian Country.* Lincoln: University of Nebraska Press.

Goffman, Erving. 1961. *Encounters: Two Studies in the Sociology of Interaction.* Indianapolis: Bobbs-Merrill.

———. 1969. *Strategic Interaction.* Philadelphia: University of Pennsylvania Press.

Goodwin, Jeff, James M. Jasper, and Francesca Polletta, editors. 2001. *Passionate Politics: Emotions and Social Movements.* Chicago: University of Chicago Press.

Gordon, Linda. 1988. *Heroes of Their Own Lives: The Politics and History of Family Violence: Boston, 1880–1960.* New York: Viking.

Grandin, Temple. 2008. *The Way I See It: A Personal Look at Autism and Asperger's.* Arlington, Texas: Future Horizons.

Grandin, Temple, and Catherine Johnson. 2005. *Animals in Translation: Using the Mysteries of Autism to Decode Animal Behavior.* New York: Scribner.

Granovetter, Mark. 1990. "The Old and the New Economic Sociology: A History and an Agenda." In *Beyond the Marketplace: Rethinking Economy and Society,* Roger Friedland and A. F. Robertson, editors. New York: Aldine de Gruyter.

Gray, Carol. 1994. *Comic Strip Conversations.* Arlington, Texas: Future Horizons.

Gray, Kurt, Adrianna C. Jenkins, Andrea S. Heberlein, and Daniel M. Wegner. 2011. "Distortions of Mind Perception in Psychopathology." *Proceedings of the National Academy of Sciences* 108: 477–79.

Halsey, Katie. 2006. "The Blush of Modesty or the Blush of Shame? Reading Jane Austen's Blushes." *Forum for Modern Language Studies* 42: 226–38.

Hamilton, Virginia. 1985. *The People Could Fly: American Black Folktales.* Leo and Diane Dillon, illustrators. New York: Alfred A. Knopf.

Hammerstein II, Oscar. 1942. *Oklahoma!.* New York: Williamson Music.

Harmgart, Heike, Steffen Huck, and Wieland Müller. 2009. "The Miracle as a Randomization Device: A Lesson from Richard Wagner's Romantic Opera *Tannhäuser und der Sängerkrieg auf Wartberg.*" *Economics Letters* 102: 33–35.

Hargreaves Heap, Shaun P., and Yanis Varoufakis. 2004. *Game Theory, Second Edition: A Critical Text.* London: Routledge.

Hart, William, Dolores Albarracín, Alice H. Eagly, Inge Brechan, Matthew J. Lindberg, and Lisa Merrill. 2009. "Feeling Validated Versus Being Correct: A Meta-Analysis of Selective Exposure to Information." *Psychological Bulletin* 135: 555–88.

Heckerling, Amy, director. 1995. *Clueless.* Paramount Pictures.

Heckerling, Amy. 2009. "The Girls Who Don't Say 'Whoo!' " In *A Truth Universally Acknowledged: 33 Great Writers on Why We Read Jane Austen,* Susannah Carson, editor. New York: Random House.

Hein, Jeanne. 1993. "Portuguese Communication with Africans on the Searoute to India." *Terrae Incognitae* 25: 41–52.

Hernandez, Raymond. 2005. "Democrats Demand Rove Apologize for 9/11 Remarks." *New York Times,* June 23.

Hess, Pamela. 2004. "Fallujah Battle Not Military's Choice." United Press International, September 13.

Hirschfeld, Lawrence, Elizabeth Bartmess, Sarah White, and Uta Frith. 2007. "Can Autistic Children Predict Behavior by Social Stereotypes?" *Current Biology* 17: R451–R452.

Hodgson, Geoffrey M. 2010. "Choice, Habit and Evolution." *Journal of Evolutionary Economics* 20: 1–18.

Hubbard, Howard. 1968. "Five Long Hot Summers and How They Grew." *Public Interest* 12: 3–24.

Hynes, William J., and William G. Doty, editors. 1993. *Mythical Trickster Figures: Contours, Contexts, and Criticisms.* Tuscaloosa: University of Alabama Press.

Ingrao, Bruna. 2001. "Economic Life in Nineteenth-Century Novels: What Economists Might Learn From Literature." In *Economics and Interdisciplinary Exchange,* Guido Erreygers, editor. London: Routledge.

Jevons, W. Stanley. 1871. *The Theory of Political Economy*. London and New York: Macmillan.

Johnson, Claudia L. 1988. *Jane Austen: Women, Politics, and the Novel*. Chicago: University of Chicago Press.

Jones, Charles C., Jr. 1888 [1969]. *Negro Myths from the Georgia Coast, Told in the Vernacular*. Boston and New York: Houghton, Mifflin.

Kaminski, Marek M. 2004. *Games Prisoners Play: The Tragicomic Worlds of Polish Prison*. Princeton: Princeton University Press.

Kelley, Robin D. G. 1993. " 'We Are Not What We Seem': Rethinking Black Working-Class Opposition in the Jim Crow South." *Journal of American History* 80: 75–112.

Kennedy, Margaret. 1950. *Jane Austen*. Denver: Alan Swallow.

Keysar, Boaz, Shuhong Lin, and Dale J. Barr. 2003. "Limits on Theory of Mind Use in Adults." *Cognition* 89: 25–41.

Khawam, René R., translator. 1980. *The Subtle Ruse: The Book of Arabic Wisdom and Guile*. London: East-West Publications.

Kimball, Jeffrey. 1998. *Nixon's Vietnam War*. Lawrence: University of Kansas Press.

Knox-Shaw, Peter. 2002. "Jane Austen's Nocturnal and Anne Finch." *English Language Notes* 39: 41–54.

———. 2004. *Jane Austen and the Enlightenment*. Cambridge: Cambridge University Press.

Kobayashi, Hiromi, and Shiro Kohshima. 2001. "Unique Morphology of the Human Eye and Its Adaptive Meaning: Comparative Studies on External Morphology of the Primate Eye." *Journal of Human Evolution* 40: 419–35.

Kraus, Michael W., Stéphane Côte, and Dacher Keltner. 2010. "Social Class, Contextualism, and Empathic Accuracy." *Psychological Science* 21: 1716–23.

Landay, Lori. 1998. *Madcaps, Screwballs, and Con Women: The Female Trickster in American Culture*. Philadelphia: University of Pennsylvania Press.

Lasker, Emanuel. 1907. *Struggle*. New York: Lasker's Publishing Company.

Leamer, Edward E. 2012. *The Craft of Economics: Lessons from the Heckscher–Ohlin Framework*. Cambridge: MIT Press.

Lebowitz, Michael A. 1988. "Is 'Analytical Marxism' Marxism?" *Science and Society* 52: 191–214.

Leeson, Robert, editor. 2000. *A.W.H. Phillips: Collected Works in Contemporary Perspective*. Cambridge: Cambridge University Press.

Legro, Jeffrey W. 1996. "Culture and Preferences in the International Cooperation Two-Step." *American Political Science Review* 90: 118–37.

Leonard, Robert J. 1995. "From Parlor Games to Social Science: von Neumann, Morgenstern, and the Creation of Game Theory 1928–1944." *Journal of Economic Literature* 33: 730–61.

———. 1997. "Value, Sign, and Social Structure: The 'Game' Metaphor and Modern Social Science." *European Journal of the History of Economic Thought* 4: 299–326.

———. 2010. *Von Neumann, Morgenstern, and the Creation of Game Theory*. Cambridge: Cambridge University Press.

Lévi-Strauss, Claude. 1963. *Structural Anthropology*. New York: Basic Books.

Levine, Lawrence W. 1977. *Black Culture and Black Consciousness*. Oxford: Oxford University Press.

Levy, David M. 2001. *How the Dismal Science Got Its Name: Classical Economics and the Ur-Text of Racial Politics*. Ann Arbor: University of Michigan Press.

Lillard, Angeline. 1998. "Ethnopsychologies: Cultural Variations in Theory of Mind." *Psychological Bulletin* 123: 3–32.

Liu, Alan. 2004. "The Humanities: A Technical Profession." American Council of Learned Societies Occasional Paper no. 63. http://www.acls.org/op63.pdf.

Livingston, Paisley. 1991. *Literature and Rationality: Ideas of Agency in Theory and Fiction*. Cambridge: Cambridge University Press.

Loewenstein, George. 2000. "Emotions in Economic Theory and Economic Behavior." *American Economic Review,* Papers and Proceedings of the One Hundred Twelfth Annual Meeting of the American Economic Association, 90: 426–32.

MacKenzie, Donald. 2006. *An Engine, Not a Camera: How Financial Models Shape Markets*. Cambridge: MIT Press.

Marshall, George, director. 1939. *You Can't Cheat an Honest Man*. Universal Pictures.

McAdam, Doug. 1983. "Tactical Innovation and the Pace of Insurgency." *American Sociological Review* 48: 735–54.

McFate, Montgomery. 2005. "The Military Utility of Understanding Adversary Culture." *Joint Force Quarterly,* Number 38, 42–48.

McKissack, Patricia C. 1986. *Flossie and the Fox*. Rachel Isadora, illustrator. New York: Dial Books for Young Readers.

Mellor, Anne K. 1993. *Romanticism and Gender*. New York: Routledge.

———. 2000. *Mothers of the Nation: Women's Political Writing in England, 1780–1830*. Bloomington: Indiana University Press.

Mill, John Stuart. 1848. *Principles of Political Economy: With Some of Their Applications to Social Philosophy*. Volume 1. London: John W. Parker.

Mohn, Klaus. 2010. "Autism in Economics? A Second Opinion." *Forum for Social Economics* 39: 191–208.

Moler, Kenneth L. 1967. "The Bennet Girls and Adam Smith on Vanity and Pride." *Philological Quarterly* 46: 567–69.

Moore, Terence. 2010. "Peyton a Double Agent? Some Think So." *Fanhouse,* February 22. http://nfl.fanhouse.com/2010/02/22/peyton-a-double-agent-some-think-so/.

Morgan, Mary S. 2010. "Models, Stories and the Economic World." *Journal of Economic Methodology* 8: 361–84.

Morgenstern, Oskar. 1928. *Wirtschaftsprognose, Eine Untersuchung ihrer Voraussetzungen und Möglichkeiten*. Vienna: J. Springer.

Morgenstern, Oskar. 1935 [1976]. "Perfect Foresight and Economic Equilibrium." Originally published as "Vollkommene Voraussicht und wirtschaftliches Gleichgewicht," *Zeitschrift für Nationalökonomie* 6: 337–57. English translation by Frank Knight, in *Selected Economic Writings of Oskar Morgenstern,* Andrew Schotter, editor. New York: New York University Press.

Morris, Errol, director. 2003. *The Fog of War: Eleven Lessons from the Life of Robert S. McNamara*. Sony Pictures Classics.

Morris, Ivor. 1987. *Mr Collins Considered: Approaches to Jane Austen*. London: Routledge and Kegan Paul.

Morrison, Toni. 1981. *Tar Baby*. New York: Alfred A. Knopf.

———. 1987. *Beloved*. New York: Alfred A. Knopf.

Most, Andrea. 1998. " 'We Know We Belong to the Land': The Theatricality of Assimilation in Rodgers and Hammerstein's *Oklahoma!*" *PMLA* 113: 77–89.

MSNBC.com. 2006. "Lasers Used on Iraqi Drivers Who Won't Stop." May 18.

Mullan, John. 2012. *What Matters in Jane Austen? Twenty Crucial Puzzles Solved*. London: Bloomsbury.

Murphy, George E. 1998. "Why Women Are Less Likely Than Men to Commit Suicide." *Comprehensive Psychiatry* 39: 165–75.

Nash, John F., Jr. 1950. "Equilibrium Points in N-Person Games." *Proceedings of the National Academy of Sciences of the United States of America* 36: 48–49.

Neil, Dan. 2008. "The (Temporarily) Old Man and the Lambo." *Los Angeles Times,* October 20.

Nelles, William. 2006. "Omniscience for Atheists: Or, Jane Austen's Infallible Narrator." *Narrative* 14: 118–31.

Nelson, Julie A. 2009. "Rationality and Humanity: A View from Feminist Economics." *Occasion: Interdisciplinary Studies in the Humanities* 1 (October 15). http://arcade.stanford.edu/journals/occasion/.

Nettle, Daniel, and Helen Clegg. 2006. "Schizotypy, Creativity and Mating Success in Humans." *Proceedings of the Royal Society B* 273: 611–15.

Nickerson, Raymond S. 1999. "How We Know—and Sometimes Misjudge—What Others Know: Imputing One's Own Knowledge to Others." *Psychological Bulletin* 125: 737–59.

Nugent, Benjamin. 2009. "The Nerds of *Pride and Prejudice*." In *A Truth Universally Acknowledged: 33 Great Writers on Why We Read Jane Austen,* Susannah Carson, editor. New York: Random House.

Oatley, Keith. 2011. "In the Minds of Others." *Scientific American Mind* 22, issue 5 (November / December).

Ober, Josiah. 2011. *Rational Cooperation in Greek Political Thought*. Manuscript, Department of Political Science, Stanford University.

O'Donoghue, Ted, and Matthew Rabin. 2001. "Choice and Procrastination." *Quarterly Journal of Economics* 116: 121–60.

Olson, Mancur. 1965. *The Logic of Collective Action: Public Goods and the Theory of Groups*. Cambridge: Harvard University Press.

O'Neill, Barry. 1982. "A Problem of Rights Arbitration from the Talmud." *Mathematical Social Sciences* 2: 345–71.

———. 1990. "The Strategy of Challenges: Two Beheading Games in Medieval Literature." Working paper, Centre for International Studies and Strategic Studies, York University.

———. 2001. "Love Tokens in the *Lai de l'Ombre*." Presented at the UCLA Conference on Political Games in the Middle Ages, March.

———. 2009. "Bargaining with a Claims Structure: Possible Solutions to a Talmudic Division Problem." Working paper, Department of Political Science, UCLA.

Palmer, Alan. 2010. *Social Minds in the Novel*. Columbus, Ohio: Ohio State University Press.

Palumbo-Liu, David. 2009. "Introduction." *Occasion: Interdisciplinary Studies in the Humanities* 1 (October 15). http://arcade.stanford.edu/journals/occasion/.

———. 2012. *The Deliverance of Others: Reading Literature in a Global Age*. Durham, N. C.: Duke University Press.

Parikh, Rohit. 2008. "Knowledge, Games and Tales from the East." Paper presented at the Indian Conference in Logic and Applications, Chennai.

Pelton, Robert D. 1980. *The Trickster in West Africa: A Study of Mythic Irony and Sacred Delight*. Berkeley and Los Angeles: University of California Press.

Pessoa, Luiz. 2008. "On the Relationship Between Emotion and Cognition." *Nature Reviews Neuroscience* 9: 148–58.

Peterson, Candida C. 2002. "Drawing Insight from Pictures: The Development of Concepts of False Drawing and False Belief in Children with Deafness, Normal Hearing, and Autism." *Child Development* 73: 1442–59.

Piggott, Patrick. 1979. *The Innocent Diversion: A Study of Music in the Life and Writings of Jane Austen*. London: Douglas Cleverdon.

Poe, Edgar Allan. 1845 [1998]. *Selected Tales*. David Van Leer, editor. Oxford: Oxford University Press.

Povinelli, Daniel J., and Jennifer Vonk. 2003. "Chimpanzee Minds: Suspiciously Human?" *Trends in Cognitive Sciences* 4: 157–60.

Pritchett, Laurie. 1985. "Interview with Laurie Pritchett." Conducted by Blackside, Inc., on November 7, 1985, for *Eyes on the Prize: America's Civil Rights Years (1954–1965)*. Washington University Libraries, Film and Media Archive, Henry Hampton Collection. http://digital.wustl.edu/e/eop/browse.html.

Pronin, Emily, and Matthew B. Kugler. 2010. "People Believe They Have More Free Will than Others." *Proceedings of the National Academy of Sciences* 107: 22469–474.

Regan, Donald. 1997. "Value, Comparability, and Choice." In *Incommensurability, Incomparability, and Practical Reason*, Ruth Chang, editor. Cambridge: Harvard University Press.

Renfroe, Anita. 2009. *Don't Say I Didn't Warn You: Kids, Carbs, and the Coming Hormonal Apocalypse*. New York: Hyperion.

Richardson, Alan. 2002. "Of Heartache and Head Injury: Reading Minds in *Persuasion*." *Poetics Today* 23: 141–60.

Robnett, Belinda. 1996. "African-American Women in the Civil Rights Movement, 1954–1965: Gender, Leadership, and Micromobilization." *American Journal of Sociology* 101: 1661–93.

Rogers, Pat, editor. 2006. Introduction and explanatory notes to *Pride and Prejudice*, by Jane Austen. Cambridge: Cambridge University Press.

Rosen, Sherwin. 1986. "The Theory of Equalizing Differences." In *Handbook of Labor Economics*, Orley Ashenfelter and Richard Layard, editors. Volume 1. Amsterdam: Elsevier Science Publishers.

Ross, Herbert, director. 1984. *Footloose.* Paramount Pictures.

Roth, Jon. 2012. "A Tribe Called Queer." *Out Magazine,* January 11.

Rubin, Alissa J., and Doyle McManus. 2004. "Why America Has Waged a Losing Battle on Fallouja." *Los Angeles Times,* October 24, 2004.

Rubinstein, Ariel. 2012a. *Economic Fables.* Cambridge: Open Book Publishers.

———. 2012b. *Lecture Notes in Microeconomic Theory: The Economic Agent.* Second edition. Princeton: Princeton University Press.

Sally, David, and Elisabeth Hill. 2006. "The Development of Interpersonal Strategy: Autism, Theory-of-Mind, Cooperation and Fairness." *Journal of Economic Psychology* 27: 73–97.

Sarin, Amita. 2005. *Akbar and Birbal.* New Delhi: Penguin.

Scahill, Jeremy. 2008. *Blackwater: The Rise of the World's Most Powerful Mercenary Army.* Revised edition. New York: Nation Books.

Schelling, Thomas C. 1960 [1980]. *The Strategy of Conflict.* Second edition. Cambridge: Harvard University Press.

Schulz, Kathryn. 2010. *Being Wrong: Adventures in the Margin of Error.* New York: HarperCollins.

Scott, James C. 1985. *Weapons of the Weak: Everyday Forms of Peasant Resistance.* New Haven: Yale University Press.

———. 1990. *Domination and the Arts of Resistance: Hidden Transcripts.* New Haven: Yale University Press.

Sedgwick, Eve Kosofsky. 1991. "Jane Austen and the Masturbating Girl." *Critical Inquiry* 17: 818–37.

Sen, Amartya K. 1967. "Isolation, Assurance, and the Social Rate of Discount." *Quarterly Journal of Economics* 81: 112–24.

Sepp, Kalev I. 2007. "From 'Shock and Awe' to 'Hearts and Minds': The Fall and Rise of US Counterinsurgency Capability in Iraq." *Third World Quarterly* 28: 217–30.

Shah, Anuj K., and Daniel M. Oppenheimer. 2008. "Heuristics Made Easy: An Effort-Reduction Framework." *Psychological Bulletin* 134: 207–22.

Shany-Ur, Tal, Pardis Poorzand, Scott N. Grossman, Matthew E. Growdon, Jung Y. Jang, Robin S. Ketelle, Bruce L. Miller, and Katherine P. Rankin. 2012. "Comprehension of Insincere Communication in Neurodegenerative Disease: Lies, Sarcasm, and Theory of Mind." *Cortex* 48: 1329–41.

Shakespeare, William. 1600 [2004]. *Much Ado About Nothing.* In *The Complete Works of Shakespeare,* David Bevington, editor. Fifth edition. New York: Pearson Longman.

Shatz, Marilyn, Gil Diesendruck, Ivelisse Martinez-Beck, and Didar Akar. 2003. "The Influence of Language and Socioeconomic Status on Children's Understanding of False Belief." *Developmental Psychology* 39: 717–29.

Shearn, Don, Erik Bergman, Katherine Hill, Andy Abel, and Lael Hinds. 1992. "Blushing as a Function of Audience Size." *Psychophysiology* 29: 431–36.

Sheinkin, Steve. 2008. *Rabbi Harvey Rides Again: A Graphic Novel of Dueling Jewish Folktales in the Wild West.* Woodstock, Vt.: Jewish Lights Publishing.

———. 2010. *Rabbi Harvey vs. The Wisdom Kid: A Graphic Novel of Dueling Jewish Folktales in the Wild West.* Woodstock, Vt.: Jewish Lights Publishing.

Sherrod, Charles. 1985. "Interview with Charles Sherrod." Conducted by Blackside, Inc., on December 20, 1985, for *Eyes on the Prize: America's Civil Rights Years (1954–1965)*. Washington University Libraries, Film and Media Archive, Henry Hampton Collection. http://digital.wustl.edu/e/eop/browse.html.

Shim, T. Youn-ja, Min-Sun Kim, and Judith N. Martin. 2008. *Changing Korea: Understanding Culture and Communication.* New York: Peter Lang.

Sihag, Balbir S. 2007. "Kautilya on Time Inconsistency Problem and Asymmetric Information." *Indian Economic Review* 42: 41–55.

Silver, Sean R. 2009/2010. "*The Rape of the Lock* and the Origins of Game Theory." *Connotations* 19: 203–28.

Silverman, Chloe. 2012. *Understanding Autism: Parents, Doctors, and the History of a Disorder.* Princeton: Princeton University Press.

Simpson, Richard. 1870 [1968]. Unsigned review of the *Memoir. North British Review*, April. Reprinted in *Jane Austen: The Critical Heritage*, B. C. Southam, editor. London: Routledge and Kegan Paul.

Singer, Tania, and Ernst Fehr. 2005. "The Neuroeconomics of Mind Reading and Empathy." *American Economic Review*, Papers and Proceedings of the One Hundred Seventeenth Annual Meeting of the American Economic Association, 95: 340–45.

Smith, Adam. 1759 [2009]. *The Theory of Moral Sentiments.* New York: Penguin.

Smith, Jeanne Rosier. 1997. *Writing Tricksters: Mythic Gambols in American Ethnic Literature.* Berkeley and Los Angeles: University of California Press.

Solomon, Robert C. 2003. *Not Passion's Slave: Emotions and Choice.* Oxford: Oxford University Press.

Southam, B. C. 1968. "Introduction." In *Jane Austen: The Critical Heritage*, B. C. Southam, editor. London: Routledge and Kegan Paul.

Stevens, Denis. 1965. Introduction to *The Art of Accompaniment from a Thorough-Bass as Practiced in the XVIIth and XVIIIth Centuries*, by F. T. Arnold. Mineola, N.Y.: Dover.

Stiglitz, Joseph E. 2010. *Freefall: America, Free Markets, and the Sinking of the World Economy.* New York: W. W. Norton.

Sun-tzu (Sunzi). 2009. *The Art of War.* Edited and translated by John Minford. New York: Penguin.

Suskind, Ron. 2004. "Without a Doubt." *New York Times Magazine*, October 17.

Sutherland, John. 1999. *Who Betrays Elizabeth Bennet? Further Puzzles in Classic Fiction.* Oxford: Oxford University Press.

Sutherland, John, and Deirdre Le Faye. 2005. *So You Think You Know Jane Austen? A Literary Quizbook.* Oxford: Oxford University Press.

Swirski, Peter. 1996. "Literary Studies and Literary Pragmatics: The Case of the 'Purloined Letter.' " *SubStance* 81: 69–89.

Taylor, Michael. 2006. *Rationality and the Ideology of Disconnection.* Cambridge: Cambridge University Press.

Thomas, James Ellis. 2000. "The Saturday Morning Car Wash Club." *The New Yorker*, July 10, 66–71.

Tobe, Keiko. 2008. *With the Light: Raising an Autistic Child [Hikari To Tomoni]*. Volume 2. Satsuki Yamashita, translator. New York: Yen Press.

Todd, Janet. 2006. *The Cambridge Introduction to Jane Austen*. Cambridge: Cambridge University Press.

Todd, Janet, and Antje Blank, editors. 2006. Explanatory notes for *Persuasion, by Jane Austen*. Cambridge: Cambridge University Press.

Tomasello, Michael, Brian Hare, Hagen Lehmann, and Josep Call. 2007. "Reliance on Head Versus Eyes in the Gaze Following of Great Apes and Human Infants: The Cooperative Eye Hypothesis." *Journal of Human Evolution* 52: 314–20.

Treitel, Guenter Heinz. 1984. "Jane Austen and the Law." *Law Quarterly Review* 100: 549–86.

Tversky, Amos, and Daniel Kahneman. 1991. "Loss Aversion in Riskless Choice: A Reference-Dependent Model." *Quarterly Journal of Economics* 106: 1039–61.

Vanderschraaf, Peter. 1998. "The Informal Game Theory in Hume's Account of Convention." *Economics and Philosophy* 14: 215-47.

Vann, David J. 1985. "Interview with David J. Vann." Conducted by Blackside, Inc., on November 1, 1985, for *Eyes on the Prize: America's Civil Rights Years (1954–1965)*. Washington University Libraries, Film and Media Archive, Henry Hampton Collection. http://digital.wustl.edu/e/eop/browse.html.

Varian, Hal R. 2006. "Revealed Preference." In *Samuelsonian Economics and the Twenty-First Century*, Michael Szenberg, Lall Ramrattan, and Aron A. Gottesman, editors. Oxford: Oxford University Press.

Vermeule, Blakey. 2010. *Why Do We Care about Literary Characters?* Baltimore: Johns Hopkins University Press.

von Neumann, John, and Oskar Morgenstern. 1944. *Theory of Games and Economic Behavior*. Princeton: Princeton University Press.

Wald, Gayle. 2000. "*Clueless* in the Neo-Colonial World Order." In *The Postcolonial Jane Austen*, You-me Park and Rajeswari Sunder Rajan, editors. London: Routledge.

Waldron, Mary. 1999. *Jane Austen and the Fiction of Her Time*. Cambridge: Cambridge University Press.

Walker, Wyatt Tee. 1985. "Interview with Wyatt Tee Walker." Conducted by Blackside, Inc., on October 11, 1985, for *Eyes on the Prize: America's Civil Rights Years (1954–1965)*. Washington University Libraries, Film and Media Archive, Henry Hampton Collection. http://digital.wustl.edu/e/eop/browse.html.

Walkowitz, Judith R. 1980. *Prostitution and Victorian Society: Women, Class, and the State*. Cambridge: Cambridge University Press.

Wallace, Robert K. 1983. *Jane Austen and Mozart: Classical Equilibrium in Fiction and Music*. Athens: University of Georgia Press.

Watts, Michael. 2002. "How Economists Use Literature and Drama." *Journal of Economic Education* 33: 377–86.

Watts, Michael, and Robert F. Smith. 1989. "Economics in Literature and Drama." *Journal of Economic Education* 20: 291–307.

White, Ralph K. 1984. *Fearful Warriors: A Psychological Profile of U.S.-Soviet Relations*. New York: Free Press.

White, Sarah, Elisabeth Hill, Joel Winston, and Uta Frith. 2006. "An Islet of Social Ability in Asperger Syndrome: Judging Social Attributes from Faces." *Brain and Cognition* 61: 69–77.

Wiese, Harald. 2012. "Backward Induction in Indian Animal Tales." *International Journal of Hindu Studies* 16: 93–103.

Williams, Juan. 1987. *Eyes on the Prize: America's Civil Rights Years, 1954–1965.* New York: Penguin.

Wiltshire, John. 1997. "*Mansfield Park, Emma, Persuasion.*" In *The Cambridge Companion to Jane Austen*, Edward Copeland and Juliet McMaster, editors. Cambridge: Cambridge University Press.

Wimmer, Heinz, and Josef Perner. 1983. "Beliefs about Beliefs: Representation and Constraining Function of Wrong Beliefs in Young Children's Understanding of Deception." *Cognition* 13: 103–28.

Woloch, Alex. 2003. *The One vs. the Many: Minor Characters and the Space of the Protagonist in the Novel.* Princeton: Princeton University Press.

Woods, Gregory. 1999. *A History of Gay Literature: The Male Tradition.* New Haven: Yale University Press.

Wright, Evan. 2004. *Generation Kill: Devil Dogs, Iceman, Captain America and The New Face of American War.* New York: G. P. Putnam's.

Wright, Richard. 1945 [1993]. *Black Boy.* New York: HarperPerennial.

Wu, Shali, and Boaz Keysar. 2007. "The Effect of Culture on Perspective Taking." *Psychological Science* 18: 600–606.

Zelazo, Philip David, and William A. Cunningham. 2007. "Executive Function: Mechanisms Underlying Emotion Regulation." In *Handbook of Emotion Regulation*, edited by James J. Gross. New York: Guilford.

Zinn, Howard. 2003. *A People's History of the United States: 1492–Present.* New York: HarperCollins.

Zunshine, Lisa. 2006. *Why We Read Fiction: Theory of Mind and the Novel.* Columbus, Ohio: Ohio State University Press.

Zunshine, Lisa. 2007. "Why Jane Austen Was Different, and Why We May Need Cognitive Science to See It." *Style* 41: 275–98.

Index

Ado Annie, 228–29
Africa, Portuguese exploration of, 216
African American folktales. *See* folktales, African American
African American labor resistance. *See* labor resistance, African American
agents, 204
aggression, 27–28, 127
aging suit, 17, 216
Akbar and Birbal, 228, 229
Albany, Georgia, 40
Ali Hakim, 228–29
Ali (son of Abu-Talib), 19
Allen, Mr., 67, 74
Allen, Mrs., 67–68, 73–74, 179, 197
  passion for dress, 189
  self-reference of, 200, 201
  as strategic sophomore, 112
  talking to herself, 192–93
  visuality of, 190
anger, 39, 51, 100, 165, 220
  and decision-making, 57, 115, 117
  of Fanny Price, 103, 208
  of General Tilney, 118, 207
animals, 16, 76, 204. *See also* horses
  in folktales, 35–39
  as model of behavior, 31
  saying people are, 225–27
Anne Elliot. *See* Elliot, Anne
apes, 18
artfulness, 137, 186, 199
  and selfishness, 133
  as strategic thinking, 108–9
artlessness, 89, 166
  and charm, 174–75
atomistic individuals, 26, 65, 141, 153
Austen, Jane
  economic hardheadedness of, 34
  empirical habits of, 72
  intellectual milieu of, 32–33
  intentions of, 175–87
  theory of value of, 230
autism
  in Austen's characters, 188–94
  and economics, 231

and maleness, 17, 196
and schizotypy, 59
and social roles, 217
and theory of mind, 16–18
Azuma, Hikaru, 217–18
Azuma, Sachiko, 217

backfiring plans, 115, 205–10, 225–27
  and strategic ineptitude, 112
backgammon, 137–38, 218, 231
Bao Ninh, 215
bargaining position, 219–24, 227
Barton Cottage, 54, 116, 164
Bates, Miss, 90–95, 130, 197, 202
  as Greek chorus, 187
  and overcontextualization, 193
Bates, Mrs., 90, 94
Bath, 120, 144, 189, 190, 192, 203
  Anne Elliot and Captain Wentworth meeting at concert in, 102, 114, 171
  Elliot family's move to, 61–67, 137, 151, 184, 199
  Sir Walter Elliot's household at, 103, 133
  Catherine Morland's trip to, 67–71
  where strategic thinking is learned, 71, 73–74, 183
Beatrice, 20–26, 159–60
beauty, 66, 78, 98, 104, 230
  estimation of, 113
  and minuteness, 189
  and weak strategic thinkers, 191, 195, 196
Benedick, 20–26, 159–60
Bennet, Elizabeth, 50–54
  as "Beatrice-like," 160
  and Jane Bennet, 171; on Caroline Bingley, 105; on Mr. Darcy, 151–52, 173; on naivety, 173, 174
  and Kitty Bennet, 53
  and Lydia Bennet, 51, 53, 104
  and Mary Bennet, 135, 191
  and Mr. Bennet, 50–51, 95, 107, 152–53
  and Mrs. Bennet, 52, 139, 146, 152
  and Caroline Bingley, 50, 112, 115, 186